2013

CRISC™ Review Questions, Answers & Explanations Manual 2013

Certified in Risk and Information Systems Control™
An ISACA® Certification

ISACA®

With more than 100,000 constituents in 180 countries, ISACA (*www.isaca.org*) is a leading global provider of knowledge, certifications, community, advocacy and education on information systems (IS) assurance and security, enterprise governance and management of IT, and IT-related risk and compliance. Founded in 1969, the nonprofit, independent ISACA hosts international conferences, publishes the *ISACA® Journal*, and develops international IS auditing and control standards, which help its constituents ensure trust in, and value from, information systems. It also advances and attests IT skills and knowledge through the globally respected Certified Information Systems Auditor® (CISA®), Certified Information Security Manager® (CISM®), Certified in the Governance of Enterprise IT® (CGEIT®) and Certified in Risk and Information Systems Control™ (CRISC™) designations.

ISACA continually updates and expands the practical guidance and product family based on the COBIT® framework. COBIT helps IT professionals and enterprise leaders fulfill their IT governance and management responsibilities, particularly in the areas of assurance, security, risk and control, and deliver value to the business.

Disclaimer

ISACA has designed and created *CRISC™ Review Questions, Answers & Explanations Manual 2013* primarily as an educational resource to assist individuals preparing to take the CRISC certification exam. It was produced independently from the CRISC exam and the CRISC Certification Committee, which has had no responsibility for its content. Copies of past exams are not released to the public and were not made available to ISACA for preparation of this publication. ISACA makes no representations or warranties whatsoever with regard to these or other ISACA publications assuring candidates' passage of the CRISC exam.

Reservation of Rights

© 2012 ISACA. All rights reserved. No part of this publication may be used, copied, reproduced, modified, distributed, displayed, stored in a retrieval system or transmitted in any form by any means (electronic, mechanical, photocopying, recording or otherwise) without the prior written authorization of ISACA.

ISACA

3701 Algonquin Road, Suite 1010
Rolling Meadows, IL 60008 USA
Phone: +1.847.253.1545
Fax: +1.847.253.1443
Email: *info@isaca.org*
Web site: *www.isaca.org*

Participate in the ISACA Knowledge Center: *www.isaca.org/knowledge-center*
Follow ISACA on Twitter: *https://twitter.com/ISACANews*
Join ISACA on LinkedIn: ISACA (Official), *http://linkd.in/ISACAOfficial*
Like ISACA on Facebook: *www.facebook.com/ISACAHQ*

ISBN 978-1-60420-327-1
CRISC™ Review Questions, Answers & Explanations Manual 2013
Printed in the United States of America

CRISC is a trademark/service mark of ISACA. The mark has been applied for or registered in countries throughout the world.

PREFACE

ISACA is pleased to offer the 200 questions in this *CRISC™ Review Questions, Answers & Explanations Manual 2013*. The purpose of this publication is to provide the CRISC candidate with sample questions and testing topics to help prepare and study for the CRISC exam.

The material in this manual consists of multiple-choice questions, answers and explanations, which are organized according to the CRISC job practice. The questions in this manual appeared in the *CRISC Review Questions, Answers & Explanations Manual 2011* and in the *CRISC Review Questions, Answers & Explanations Manual 2012 Supplement*. These questions, answers and explanations are intended to introduce CRISC candidates to the types of questions that may appear on the CRISC exam. They are not actual questions from the exam. Questions are sorted by CRISC job practice domains and a sample exam of 200 questions is also provided. Sample questions contained in this manual are provided to assist the CRISC candidate in understanding the material in the *CRISC™ Review Manual 2013* and to depict the type of question format typically found on the CRISC exam.

Some of the questions are presented in scenarios. Scenarios are mini-case studies that describe a situation or an enterprise and require candidates to answer one or more questions based on the information provided. A scenario can focus on one or more domains. The CRISC exam includes scenario questions.

ISACA wishes you success with the CRISC exam and welcomes your comments and suggestions on the use and coverage of this manual. Once you have completed your exam, please take a moment to complete the online evaluation that corresponds to this publication (*www.isaca.org/studyaidsevaluation*). Your observations will be invaluable as new questions, answers and explanations are prepared.

ACKNOWLEDGMENTS

This *CRISC™ Review Questions, Answers & Explanations Manual 2013* is the result of the collective efforts of many volunteers. ISACA members from throughout the world participated, generously offering their talent and expertise. This international team exhibited a spirit and selflessness that has become the hallmark of contributors to this valuable manual. Their participation and insight are truly appreciated.

TABLE OF CONTENTS

PREFACE iii

ACKNOWLEDGMENTS iv

INTRODUCTION vii
    DOCUMENT STRUCTURE vii
    CRISC EXAM QUESTIONS viii

PRETEST ix

QUESTIONS, ANSWERS AND EXPLANATIONS BY DOMAIN 1
    DOMAIN 1—RISK IDENTIFICATION, ASSESSMENT AND EVALUATION (31%) 1
    DOMAIN 2—RISK RESPONSE (17%) 29
    DOMAIN 3—RISK MONITORING (17%) 45
    DOMAIN 4—INFORMATION SYSTEMS CONTROL DESIGN AND IMPLEMENTATION (17%) 61
    DOMAIN 5—INFORMATION SYSTEMS CONTROL MONITORING AND MAINTENANCE (18%) 75

POSTTEST 91

SAMPLE EXAM 93

SAMPLE EXAM ANSWER AND REFERENCE KEY 123

SAMPLE EXAM ANSWER SHEET (PRETEST) 125

SAMPLE EXAM ANSWER SHEET (POSTTEST) 127

EVALUATION 129

PREPARE FOR THE 2013 CRISC EXAMS 130

Page intentionally left blank

INTRODUCTION

The *CRISC™ Review Questions, Answers & Explanations Manual 2013* has been developed to assist the CRISC candidate in studying and preparing for the CRISC exam. As you use this publication to prepare for the exam, please note that the exam covers a broad spectrum of IS control solutions and how they relate to business and IT risk management issues. Do not assume that reading and working the questions in this manual will fully prepare you for the exam. Since exam questions often relate to practical experience, CRISC candidates are advised to refer to their own experience and to other publications and frameworks referred to in the *CRISC™ Review Manual 2013*, such as *The Risk IT Practitioner Guide* and the Risk IT and COBIT frameworks. These additional references are excellent sources of further detailed information and clarification. It is suggested that candidates evaluate the domains in which they feel weak or require a further understanding and then study accordingly.

Also, please note that this publication has been written using standard American English. This publication is ideal to use in conjunction with the *CRISC™ Review Manual 2013*. The questions in this publication are not actual CRISC exam questions, but are intended to provide the CRISC candidate with an understanding of the type and structure of questions that have typically appeared on the exam.

Document Structure

This manual consists of 200 sample multiple-choice questions, answers and explanations. These questions are provided in two formats:
1. Questions Sorted by Domain
2. Sample Exam

1. Questions Sorted by Domain

Questions, answers and explanations are sorted by CRISC domain and contain the number of items equivalent to the percentages indicated in the 2013 CRISC job practice:

Domain 1—Risk Identification, Assessment and Evaluation .... 31 percent
Domain 2—Risk Response .... 17 percent
Domain 3—Risk Monitoring .... 17 percent
Domain 4—Information Systems Control Design and Implementation .... 17 percent
Domain 5—Information Systems Control Monitoring and Maintenance .... 18 percent

Scenarios
Some of the questions are presented in scenarios. Scenarios are mini-case studies that describe a situation or an enterprise and require candidates to answer one or more questions based on the information provided. A scenario can focus on one or more domains. The CRISC exam may include scenario questions.

2. Sample Exam
The 200 questions have been provided in a sample exam in random order.

Candidates are urged to use this sample exam and the answer sheets provided in this publication to simulate an actual exam. Many candidates use this sample exam as a pretest to determine their specific strengths or weaknesses, or as a final test to determine their readiness to sit for the exam. Sample exam answer sheets have been provided for both uses, and an answer/reference key is included. This sample exam has been cross-referenced to the questions, answers and explanations by domain so that it is convenient to refer back to the explanations of the correct answers.

INTRODUCTION

CRISC Exam Questions

Background
The CRISC candidate should recognize that individual perceptions and experiences may not reflect the more global position or circumstance. Since the CRISC exam and manuals are written from a global perspective, the candidate will be required to be somewhat flexible when reading about a condition that may be contrary to the candidate's experience. It should be noted that actual CRISC exam questions are written by experienced IS risk and control practitioners from around the world. Each question on the exam is reviewed by ISACA's CRISC Test Enhancement Subcommittee and ISACA's CRISC Certification Committee, both of which consist of international members. This geographic representation ensures that all test questions will be understood equally in each country and language.

Purpose and Structure
CRISC exam questions are developed with the intent of measuring and testing practical knowledge and applying general concepts and standards.

Each CRISC question has a stem (question) and four choices (answers). The candidate is asked to choose the correct or best answer from the choices. The stem may be in the form of a question or an incomplete statement. In some instances, a scenario or description problem may also be included. These questions normally include a description of a situation and require the candidate to answer two or more questions based on the information provided. Please note that questions requiring the candidate to choose one to several items from a list are no longer used on the CRISC exam and should not be used as a study source. All questions are presented in a multiple-choice format and are designed for one best answer.

How to Find the Correct Answer
The candidate is cautioned to read each question carefully. Many times, a CRISC exam question will require the candidate to choose the appropriate answer that is **MOST** likely or **BEST**. Other times, a candidate may be asked to choose a practice or procedure that would be performed **FIRST** related to the other choices. In every case, the candidate is required to read the question carefully, eliminate known wrong choices and then make the best choice possible. Knowing the types of questions asked on the exam and how to study to correctly answer them will assist the CRISC candidate in successfully preparing for the CRISC exam.

> **Note:** ISACA review manuals are living documents. As technology advances, ISACA manuals will be updated to reflect such advances. Further updates or corrections to this document before the date of the exam may be viewed at *www.isaca.org/studyaidupdates*.

PRETEST

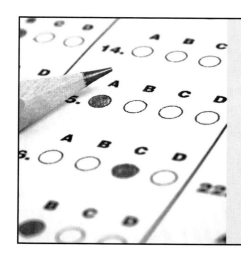

If you wish to take a pretest to determine strengths and weaknesses, the Sample Exam begins on page 93 and the pretest answer sheet begins on page 125. You can score your pretest with the Sample Exam Answer and Reference Key on page 123.

DOMAIN 1—RISK IDENTIFICATION, ASSESSMENT AND EVALUATION

QUESTIONS, ANSWERS AND EXPLANATIONS BY DOMAIN

DOMAIN 1—RISK IDENTIFICATION, ASSESSMENT AND EVALUATION (31%)

R1-1 Which of the following uses risk scenarios when estimating the likelihood and impact of significant risk to the organization?

A. An IT audit
B. A security gap analysis
C. A threat and vulnerability assessment
D. An IT security assessment

C is the correct answer.

Justification:
A. An IT audit typically uses technical evaluation tools or assessment methodologies to enumerate risk; generally, this is done for the purpose of prioritizing audit projects or for delineating the scope of an audit.
B. A security gap analysis typically uses technical evaluation tools or assessment methodologies to enumerate risk or areas of noncompliance, but does not utilize risk scenarios.
C. A threat and vulnerability assessment typically evaluates all elements of a business process for threats and vulnerabilities and identifies the likelihood of occurrence and the business impact if the threats were realized.
D. An IT security assessment typically uses technical evaluation tools or assessment methodologies to enumerate risk or areas of noncompliance, but does not utilize risk scenarios.

R1-2 Which of the following is **MOST** important to determine when defining risk management strategies?

A. Risk assessment criteria
B. IT architecture complexity
C. An enterprise disaster recovery plan (DRP)
D. Organizational objectives and risk tolerance

D is the correct answer.

Justification:
A. Risk assessment criteria become part of this framework, but only after proper analysis.
B. IT architecture complexity is more directly related to assessing risk than defining strategies.
C. An enterprise DRP is more directly related to assessing risk than defining strategies.
D. While defining risk management strategies, the risk practitioner needs to analyze the organization's objectives and risk tolerance and define a risk management framework based on this analysis. Some organizations may accept known risk, while others may invest in and apply mitigating controls to reduce risk.

DOMAIN 1—RISK IDENTIFICATION, ASSESSMENT AND EVALUATION

R1-3 Which of the following is the **BEST** reason to perform a risk assessment?

A. To satisfy regulatory requirements
B. To budget appropriately for needed controls
C. To analyze the effect on the business
D. To help determine the current state of risk

D is the correct answer.

Justification:
A. Performing a risk assessment may satisfy regulatory requirements, but is not the reason to perform a risk assessment.
B. Budgeting appropriately may come as a result, but is not the reason to perform a risk assessment.
C. Analyzing the effect on the business is part of the process, but the needs or acceptable effect or response must also be determined.
D. **The risk assessment is used to identify and evaluate the impact of failure on critical business processes (and IT components supporting them) and to determine time frames, priorities, resources and interdependencies. It is part of the process to help determine the current state of risk and helps determine risk countermeasures in alignment with business objectives.**

R1-4 Which of the following is the **MOST** important information to include in a risk management strategic plan?

A. Risk management staffing requirements
B. The risk management mission statement
C. Risk mitigation investment plans
D. The current state and desired future state

D is the correct answer.

Justification:
A. Risk management staffing requirements are generally driven by a robust understanding of the current and desired future state.
B. The risk management mission statement is important, but is not an actionable part of a risk management strategic plan.
C. Risk mitigation investment plans are generally driven by a robust understanding of the current and desired future state.
D. **It is most important to paint a vision for the future and then draw a road map from the starting point; therefore, this requires that the current state and desired future state be fully understood.**

DOMAIN 1—RISK IDENTIFICATION, ASSESSMENT AND EVALUATION

R1-5 Which of the following is **MOST** important when evaluating and assessing risk to an enterprise or business process?

A. Identification of controls that are currently in place to mitigate identified risk
B. Threat intelligence, including likelihood of identified threats
C. Historical risk assessment data
D. Control testing results

B is the correct answer.

Justification:
A. Identification of controls that are currently in place is an important part of the risk assessment process, but is not as important as threat intelligence.
B. **One of the key requirements of effective risk assessment is its association and alignment with current intelligence that includes data on the likelihood of identified threats. The probability of risk being realized is one of the primary determinations of risk prioritization.**
C. Historical risk assessment data are useful in understanding previously identified risk, but are not essential to the risk assessment process.
D. Control testing results are a component of risk assessment that helps support conclusions. Threat intelligence will often drive the testing of specific controls based on the identification of risk scenarios during the evaluation and assessment activity. These data are valuable to the risk assessment process, but are not as valuable as accurate threat intelligence.

R1-6 Information that is no longer required to support the main purpose of the business from an information security perspective should be:

A. analyzed under the retention policy.
B. protected under the information classification policy.
C. analyzed under the backup policy.
D. protected under the business impact analysis (BIA).

A is the correct answer.

Justification:
A. **Information that is no longer required should be analyzed under the retention policy to determine whether the organization is required to maintain the data for business, legal or regulatory reasons. Keeping data that are no longer required unnecessarily consumes resources; may be in breach of legal and regulatory obligations regarding retention of data; and, in the case of sensitive personal information, can increase the risk of data compromise.**
B. The information classification policy should specify retention and destruction of information that is no longer of value to the core business, as applicable.
C. The backup policy is generally based on recovery point objectives (RPOs). The information classification policy should specify retention and destruction of backup media.
D. A BIA can help determine that this information does not support the main objective of the business, but does not indicate the action to take.

DOMAIN 1—RISK IDENTIFICATION, ASSESSMENT AND EVALUATION

R1-7 An enterprise has outsourced the majority of its IT department to a third party whose servers are in a foreign country. Which of the following is the **MOST** critical security consideration?

A. A security breach notification may get delayed due to the time difference.
B. Additional network intrusion detection sensors should be installed, resulting in additional cost.
C. The enterprise could be unable to monitor compliance with its internal security and privacy guidelines.
D. Laws and regulations of the country of origin may not be enforceable in the foreign country.

D is the correct answer.

Justification:
A. Security breach notification is not a problem. Time difference does not play a role in a 24/7 environment. Pagers, cellular telephones, etc., are usually available to communicate a notification.
B. The need for additional network intrusion sensors is a manageable problem that requires additional funding, but can be addressed.
C. Outsourcing does not remove the enterprise's responsibility regarding internal requirements.
D. **Laws and regulations of the country of origin may not be enforceable in the foreign country. Conversely, the laws and regulations of the foreign outsourcer may also impact the enterprise. A potential violation of local laws applicable to the enterprise or the vendor may not be recognized or rectified due to the lack of knowledge of the local laws that are applicable and the inability to enforce those laws.**

R1-8 Which of the following will have the **MOST** significant impact on standard information security governance models?

A. Number of employees
B. Cultural differences between physical locations
C. Complexity of the organizational structure
D. Evolving legislative requirements

C is the correct answer.

Justification:
A. The number of employees has less impact on information security governance models because well-defined process, technology and personnel components intermingle to provide the proper governance.
B. The distance between physical locations has less impact on information security governance models because well-defined process, technology and personnel components intermingle to provide the proper governance.
C. **Information security governance models are highly dependent on the complexity of the organizational structure. Some of the elements that impact organizational structure are multiple business units and functions across the organization, leadership and lines of communication.**
D. Being current with changing legislative requirements should not have a major impact once good governance models are in place; therefore, governance will help in effective management of the organization's ongoing compliance.

DOMAIN 1—RISK IDENTIFICATION, ASSESSMENT AND EVALUATION

R1-9 Which of the following would data owners be **PRIMARILY** responsible for when establishing risk mitigation methods?

A. Intrusion detection
B. Antivirus controls
C. User entitlement changes
D. Platform security

C is the correct answer.

Justification:
A. Data custodians are responsible for designing and implementing intrusion detection, based on business needs.
B. Data custodians are responsible for designing and implementing antivirus controls, based on business needs.
C. **Data owners are responsible for assigning user entitlement changes and approving access to the systems for which they are responsible.**
D. Data custodians are responsible for designing and implementing platform security, based on business needs.

R1-10 An enterprise recently developed a breakthrough technology that could provide a significant competitive edge. Which of the following **FIRST** governs how this information is to be protected from within the enterprise?

A. The data classification policy
B. The acceptable use policy
C. Encryption standards
D. The access control policy

A is the correct answer.

Justification:
A. **A data classification policy describes the data classification categories; level of protection to be provided for each category of data; and roles and responsibilities of potential users, including data owners.**
B. An acceptable use policy is oriented more toward the end user and, therefore, does not specifically address which controls should be in place to adequately protect information.
C. Mandated levels of protection, as defined by the data classification policy, should drive which levels of encryption will be in place.
D. Mandated levels of protection, as defined by the data classification policy, should drive which access controls will be in place.

DOMAIN 1—RISK IDENTIFICATION, ASSESSMENT AND EVALUATION

R1-11 Malware has been detected that redirects users' computers to web sites crafted specifically for the purpose of fraud. The malware changes domain name system (DNS) server settings, redirecting users to sites under the hackers' control. This scenario **BEST** describes a:

A. man-in-the-middle (MITM) attack.
B. phishing attack.
C. pharming attack.
D. social engineering attack.

C is the correct answer.

Justification:
A. In an MITM attack, the attacker intercepts the communication stream between two parts of the victim system and then replaces the traffic between the two components with the intruder's own, eventually assuming control of the communication.
B. A phishing attack is a type of electronic mail (email) attack that attempts to convince a user that the originator is genuine, but with the intention of obtaining information for use in social engineering.
C. **A pharming attack is an MITM attack that changes the pointers on a DNS server and redirects a user's session to a masquerading web site.**
D. A social engineering attack deceives users or administrators at the target site into revealing confidential or sensitive information. They can be executed person-to-person, over the telephone or via email.

R1-12 How often should risk be evaluated?

A. Annually or when there is a significant change
B. Once a year for each business process and subprocess
C. Every three to six months for critical business processes
D. Only after significant changes occur

A is the correct answer.

Justification:
A. **Risk is constantly changing. Evaluating risk annually or when there is a significant change offers the best alternative because it takes into consideration a reasonable time frame while allowing flexibility to address significant change.**
B. Evaluating risk once a year is insufficient if important changes take place.
C. Evaluating risk every three to six months for critical processes may not be necessary or it may not address important changes in a timely manner.
D. Evaluating risk only after significant changes occur may not take into consideration the effect of time and less significant changes that may collectively affect the overall risk

DOMAIN 1—RISK IDENTIFICATION, ASSESSMENT AND EVALUATION

R1-13 It is **MOST** important for a risk evaluation to:

A. take into account the potential size and likelihood of a loss.
B. consider inherent and control risk.
C. include a benchmark of similar companies in its scope.
D. assume an equal degree of protection for all assets.

A is the correct answer.

Justification:
A. **Risk evaluation should take into account the potential size and likelihood of a loss.**
B. Although inherent and control risk should be considered in the analysis, the impact of the risk (potential likelihood and impact of loss) should be the primary driver.
C. Risk evaluation can include comparisons with a group of companies of similar size.
D. Risk evaluation should not assume an equal degree of protection for all assets because assets may have different risk factors.

R1-14 What is the **MOST** effective method to evaluate the potential impact of legal, regulatory and contractual requirements on business objectives?

A. A compliance-oriented gap analysis
B. Interviews with business process stakeholders
C. A mapping of compliance requirements to policies and procedures
D. A compliance-oriented business impact analysis (BIA)

D is the correct answer.

Justification:
A. A gap analysis will only identify the gaps in compliance to current requirements and will not identify impacts to business objectives or activities.
B. Interviews with key business process stakeholders will identify business objectives, but will not necessarily account for the compliance requirements that must be met.
C. Mapping requirements to policies and procedures will identify how compliance is being achieved, but will not identify business impact.
D. **A compliance-oriented BIA will identify all of the compliance requirements to which the enterprise has to align and their impacts on business objectives and activities.**

DOMAIN 1—RISK IDENTIFICATION, ASSESSMENT AND EVALUATION

R1-15 Which of the following is the **BEST** way to ensure that an accurate risk register is maintained over time?

A. Monitor key risk indicators (KRIs), and record the findings in the risk register.
B. Publish the risk register centrally with workflow features that periodically poll risk assessors.
C. Distribute the risk register to business process owners for review and updating.
D. Utilize audit personnel to perform regular audits and to maintain the risk register.

B is the correct answer.

Justification:
A. Monitoring KRIs will only provide insights to known and identified risk and will not account for risk that has yet to be identified.
B. **Centrally publishing the risk register and enabling periodic polling of risk assessors through workflow features will ensure accuracy of content. A knowledge management platform with workflow and polling features will automate the process of maintaining the risk register.**
C. Business process owners typically cannot effectively identify risk to their business processes. They may not have the ability to be unbiased in their review and may not have the appropriate skills or tools to effectively evaluate risk.
D. Audit personnel may not have the appropriate business knowledge or training in risk assessment to appropriately identify risk. Regular audits of business processes can also be a hindrance to business activities and will most likely not be allowed by business leadership.

R1-16 Shortly after performing the annual review and revision of corporate policies, a risk practitioner becomes aware that a new law may affect security requirements for the human resources system. The risk practitioner should:

A. analyze in detail how the law may affect the enterprise.
B. ensure necessary adjustments are implemented during the next review cycle.
C. initiate an *ad hoc* revision of the corporate policy.
D. notify the system custodian to implement changes.

A is the correct answer.

Justification:
A. **Assessing how the law may affect the enterprise is the best course of action. The analysis must also determine whether existing controls already address the new requirements.**
B. Ensuring necessary adjustments are implemented during the next review cycle is not the best answer, particularly in cases where the law does affect the enterprise. While an annual review cycle may be sufficient in general, significant changes in the internal or external environment should trigger an *ad hoc* reassessment.
C. Initiating an *ad hoc* amendment to the corporate policy may be a rash and unnecessary action.
D. Notifying the system custodian to implement changes is inappropriate. Changes to the system should be implemented only after approval by the process owner.

DOMAIN 1—RISK IDENTIFICATION, ASSESSMENT AND EVALUATION

R1-17 Which of the following will produce comprehensive results when performing a qualitative risk analysis?

A. A vulnerability assessment
B. Scenarios with threats and impacts
C. The value of information assets
D. Estimated productivity losses

B is the correct answer.

Justification:
A. A vulnerability assessment itself provides a one-sided view unless it is linked to specific risk scenarios that help determine likelihood and impact.
B. **Using a list of possible scenarios with threats and impacts will better frame the range of risk and facilitate a more informed discussion and decision.**
C. The value of information assets is an important starting point when performing a qualitative risk analysis. However, value without consideration of realistic threats and determination of likelihood and impact is not sufficient for a risk analysis.
D. Estimated productivity losses may be an important input into the projected magnitude of an impact. However, this choice is insufficient on its own.

R1-18 When performing a risk assessment on the impact of losing a server, calculating the monetary value of the server should be based on the:

A. cost to obtain a replacement.
B. annual loss expectancy (ALE).
C. cost of the software stored.
D. original cost to acquire.

A is the correct answer.

Justification:
A. **The value of the server should be based on its replacement cost; however, the financial impact to the enterprise may be much broader, based on the function that the server performs for the business and the value it brings to the enterprise.**
B. The ALE for all risk related to the server does not represent the server's value.
C. The software can be restored from backup media.
D. The original cost may be significantly different from the current cost and, therefore, not as relevant.

R1-19 Which of the following factors should be included when assessing the impact of losing network connectivity for 18 to 24 hours?

A. The hourly billing rate charged by the carrier
B. Financial losses incurred by affected business units
C. The value of the data transmitted over the network
D. An aggregate compensation of all affected business users

B is the correct answer.

Justification:
A. The hourly billing rate charged by the carrier may be a factor that contributes to the overall financial impact; however, it is a very limited subset of the actual impact of losing network connectivity.
B. **The impact of network unavailability is the cost it incurs to the enterprise.**
C. The value of the data transmitted over the network is a subset of the financial losses incurred by affected business units.
D. An aggregate compensation of all affected business users is a subset of the financial losses incurred by affected business units.

DOMAIN 1—RISK IDENTIFICATION, ASSESSMENT AND EVALUATION

R1-20 An objective of a risk management program is to:

A. maintain residual risk at an acceptable level.
B. implement preventive controls for every threat.
C. remove all inherent risk.
D. reduce inherent risk to zero.

A is the correct answer.

Justification:
A. **Ensuring that all residual risk is maintained at a level acceptable to the business is the objective of a risk management program.**
B. The objective of a risk management program is not to implement controls for every threat.
C. A risk management program is not intended to remove every identified risk.
D. Inherent risk—the risk level of an activity, business process, or entity without taking into account the actions that management has taken or may take—is always greater than zero.

R1-21 Assessing information systems risk is **BEST** achieved by:

A. using the enterprise's past actual loss experience to determine current exposure.
B. reviewing published loss statistics from comparable organizations.
C. evaluating threats associated with existing information systems assets and information systems projects.
D. reviewing information systems control weaknesses identified in audit reports.

C is the correct answer.

Justification:
A. Past actual loss experience is a potentially useful input to the risk assessment process, but it does not address realistic risk scenarios that have not occurred in the past.
B. Published loss statistics from comparable organizations is a potentially useful input to the risk assessment process, but does not address enterprise specific risk scenarios or those that have not occurred in the past.
C. **To assess IT risk, threats and vulnerabilities need to be evaluated using qualitative or quantitative risk assessment approaches.**
D. Control weaknesses and other vulnerabilities are an important input to the risk assessment process, but by themselves are not useful.

R1-22 Which of the following is the **MOST** important requirement for setting up an information security infrastructure for a new system?

A. Performing a business impact analysis (BIA)
B. Considering personal devices as part of the security policy
C. Basing the information security infrastructure on a risk assessment
D. Initiating IT security training and familiarization

C is the correct answer.

Justification:
A. Typically, a BIA is carried out to prioritize business processes as part of a business continuity plan (BCP).
B. While considering personal devices as part of the security policy may be a consideration, it is not the most important requirement.
C. **The information security infrastructure should be based on a risk assessment.**
D. Initiating IT security training may not be important for the purpose of the information security infrastructure.

DOMAIN 1—RISK IDENTIFICATION, ASSESSMENT AND EVALUATION

R1-23 The **PRIMARY** concern of a risk practitioner documenting a formal data retention policy is:

A. storage availability.
B. applicable organizational standards.
C. generally accepted industry best practices.
D. business requirements.

D is the correct answer.

Justification:
A. Storage is irrelevant because whatever is needed must be provided.
B. Applicable organizational standards support the policy, but do not dictate it.
C. Best practices may be a useful guide, but not a primary concern.
D. **The primary concern is business requirements.**

R1-24 Which of the following environments typically represents the **GREATEST** risk to organizational security?

A. An enterprise data warehouse
B. A load-balanced, web server cluster
C. A centrally managed data switch
D. A locally managed file server

D is the correct answer.

Justification:
A. Enterprise data warehouses are generally subject to close scrutiny, good change control practices and monitoring.
B. Load-balanced, web server clusters are generally subject to close scrutiny, good change control practices and monitoring.
C. Centrally managed data switches are generally subject to close scrutiny, good change control practices and monitoring.
D. **A locally managed file server will be the least likely to conform to organizational security policies because it is generally subject to less oversight and monitoring.**

R1-25 Which of the following areas is **MOST** susceptible to the introduction of an information-security-related vulnerability?

A. Tape backup management
B. Database management
C. Configuration management
D. Incident response management

C is the correct answer.

Justification:
A. Tape backup management is generally less susceptible to misconfiguration issues than configuration management.
B. Database management is generally less susceptible to misconfiguration issues than configuration management.
C. **Configuration management provides the greatest likelihood of information security weaknesses through misconfiguration and failure to update operating system (OS) code correctly and on a timely basis.**
D. Incident response management is generally less susceptible to misconfiguration issues than configuration management.

DOMAIN 1—RISK IDENTIFICATION, ASSESSMENT AND EVALUATION

R1-26 Which of the following is the **BEST** method to analyze risk, incidents and related interdependencies to determine the impact on organizational goals?

A. Security information and event management (SIEM) solutions
B. A business impact analysis (BIA)
C. Enterprise risk management (ERM) steering committee meetings
D. Interviews with business leaders to develop a risk profile

B is the correct answer.

Justification:
A. SIEM solutions will primarily account for technical risk and typically do not evaluate the impact that business process objectives have on operational components.
B. **A BIA should include the examination of risk, incidents and interdependencies as part of the activity to identify impact to business objectives.**
C. ERM steering committees are useful for reviewing analyses that have been completed, but not for conducting analysis activities.
D. Interviews with business leaders will assist in identifying risk tolerance and key business objectives and activities, but will not analyze risk or incidents.

R1-27 Which of the following is the **MOST** important reason for conducting security awareness programs throughout an enterprise?

A. Reducing the risk of a social engineering attack
B. Training personnel in security incident response
C. Informing business units about the security strategy
D. Maintaining evidence of training records to ensure compliance

A is the correct answer.

Justification:
A. **Social engineering is the act of manipulating people into divulging confidential information or performing actions that allow an unauthorized individual to get access to sensitive information and/or systems. People are often considered the weakest link in security implementations and security awareness would help reduce the risk of successful social engineering attacks by informing and sensitizing employees about various security policies and security topics, thus ensuring compliance from each individual.**
B. Training individuals in security incident response targets is a corrective control action and not as important as proactively preventing an incident.
C. Informing business units about the security strategy is best done through steering committee meetings or other forums.
D. Maintaining evidence of training records to ensure compliance is an administrative, documentary task, but should not be the objective of training.

DOMAIN 1—RISK IDENTIFICATION, ASSESSMENT AND EVALUATION

R1-28 Overall business risk for a particular threat can be expressed as the:

A. magnitude of the impact should a threat source successfully exploit the vulnerability.
B. likelihood of a given threat source exploiting a given vulnerability.
C. product of the probability and magnitude of the impact if a threat exploits a vulnerability.
D. collective judgment of the risk assessment team.

C is the correct answer.

Justification:
A. The magnitude of the impact of a successful threat provides only one factor.
B. The likelihood alone of the impact of a successful threat provides only one factor.
C. The product of the probability and magnitude of the impact provides the best measure of the risk to an asset.
D. The judgment of the risk assessment team defines the risk on an arbitrary basis and is not suitable for a scientific risk management process.

R1-29 When developing risk scenarios for an enterprise, which of the following is the **BEST** approach?

A. The top-down approach for capital-intensive enterprises
B. The top-down approach because it achieves automatic buy-in
C. The bottom-up approach for unionized enterprises
D. The top-down and the bottom-up approach because they are complementary

D is the correct answer.

Justification:
A. Both risk scenario development approaches should be considered simultaneously, regardless of the industry.
B. Both risk scenario development approaches should be considered simultaneously, regardless of the risk appetite.
C. Both risk scenario development approaches should be considered simultaneously, regardless of the industry.
D. The top-down and bottom-up risk scenario development approaches are complementary and should be used simultaneously. In a top-down approach, one starts from the overall business objectives and performs an analysis of the most relevant and probable risk scenarios impacting the business objectives. In a bottom-down approach, a list of generic risk scenarios is used to define a set of more concrete and customized scenarios, applied to the individual enterprise's situation.

R1-30 Which of the following is the **GREATEST** risk of a policy that inadequately defines data and system ownership?

A. Audit recommendations may not be implemented.
B. Users may have unauthorized access to originate, modify or delete data.
C. User management coordination does not exist.
D. Specific user accountability cannot be established.

B is the correct answer.

Justification:
A. A policy that inadequately defines data and system ownership generally does not affect the implementation of audit recommendations, particularly because audit reports assign remediation owners.
B. Without a policy defining who has the responsibility for granting access to specific data or systems, there is an increased risk that one could gain (be given) system access without a justified business need. There is a better chance that business objectives will be properly supported when authority to grant access is assigned to specific individuals.
C. While a policy that inadequately defines data and system ownership may affect user management coordination, the greatest risk would be inappropriate granting of user access.
D. User accountability is established by assigning unique user IDs and tracking transactions.

DOMAIN 1—RISK IDENTIFICATION, ASSESSMENT AND EVALUATION

R1-31 A lack of adequate controls represents:

A. a vulnerability.
B. an impact.
C. an asset.
D. a threat.

A is the correct answer.

Justification:
A. **The lack of adequate controls represents a vulnerability, exposing sensitive information and data to the risk of malicious damage, attack or unauthorized access by hackers. This could result in a loss of sensitive information, financial loss, legal penalties, etc.**
B. Impact is the measure of the financial loss that a threat event may have.
C. An asset is something of either tangible or intangible value worth protecting, including people, systems, infrastructure, finances and reputation.
D. A threat is a potential cause of an unwanted incident.

R1-32 A risk assessment process that uses likelihood and impact in calculating the level of risk is a:

A. qualitative process.
B. failure modes and effects analysis (FMEA).
C. fault tree analysis.
D. quantitative process.

D is the correct answer.

Justification:
A. A qualitative risk assessment process uses scenarios and ranking of risk levels in calculating the level of risk.
B. An FMEA determines the extended impact of an adverse event on other systems or operational areas.
C. A fault tree analysis risk assessment determines threats by considering all threat sources to a business process.
D. **A quantitative risk assessment process uses likelihood and impact in calculating the monetary value of risk.**

R1-33 Who should be accountable for the risk to an IT system that supports a critical business process?

A. IT management
B. Senior management
C. The risk management department
D. System users

B is the correct answer.

Justification:
A. IT management is responsible for managing information systems on behalf of the business owners; they are not accountable for the risk.
B. **The accountable party is senior management. While they may not be responsible for executing the risk management program, they are ultimately liable for the acceptance and mitigation of all risk.**
C. The risk management department is responsible for the execution of the risk management program and will identify, evaluate and report on risk and risk response efforts; the department is not accountable for the risk.
D. System users are responsible for using the system properly and following procedures; they are not accountable for the risk.

DOMAIN 1—RISK IDENTIFICATION, ASSESSMENT AND EVALUATION

R1-34 Which of the following practices **BEST** mitigates the risk associated with outsourcing a business function?

A. Performing audits to verify compliance with contract requirements
B. Requiring all vendor staff to attend annual awareness training sessions
C. Retaining copies of all sensitive data on internal systems
D. Reviewing the financial records of the vendor to verify financial soundness

A is the correct answer.

Justification:
A. **Regular audits verify that the vendor is compliant with contract requirements.**
B. Requiring the vendor staff to attend annual awareness sessions is not usually part of an outsourcing contract, although it may be a good idea.
C. Keeping copies of all sensitive data is an unnecessary expenditure and may result in errors or inconsistencies with data stored at the vendor site. In addition, duplicating sensitive data means that the company is now liable for protecting data in two or more locations and increases the possibility of inappropriate access and/or data leakage.
D. Although it is common practice to review financial solvency before selecting a vendor to ensure that the vendor functions without the threat of liquidation for the foreseeable future, reviewing solvency is not the best practice to address the risk related to outsourcing an IT or business function.

R1-35 Which of the following **BEST** describes the role of management in implementing a risk management strategy?

A. Ensure that the planning, budgeting and performance of information security components are appropriate.
B. Assess and incorporate the results of the risk management activity into the decision-making process.
C. Identify, evaluate and minimize risk to IT systems that support the mission of the organization.
D. Understand the risk management process so that appropriate training materials and programs can be developed.

B is the correct answer.

Justification:
A. Ensuring the planning, budgeting and performance of information security components is usually the responsibility of the chief information officer (CIO). Although the CIO is a member of senior management, this does not best describe the collective role of senior management in establishing and implementing the risk management strategy.
B. **Assessing and incorporating the results of the risk management activity into the decision-making process best describes the role of senior management in establishing and implementing a risk management strategy.**
C. Identifying, evaluating and minimizing risk to IT systems supporting the corporate mission is done by IT security managers or an IT security function, but this does not best describe the role of senior management in creating the risk management strategy.
D. Understanding the risk management process to develop appropriate training materials and programs is usually the role of corporate security trainers, not of senior management.

DOMAIN 1—RISK IDENTIFICATION, ASSESSMENT AND EVALUATION

R1-36 Which of the following risk assessment outputs is **MOST** suitable to help justify an organizational information security program?

A. An inventory of risk that may impact the enterprise
B. Documented threats to the enterprise
C. Evaluation of the consequences
D. A list of appropriate controls for addressing risk

D is the correct answer.

Justification:
A. A risk inventory is not the best choice because it does not sufficiently address how the risk will be addressed.
B. Documentation of threats is not the best choice because it does not sufficiently address how the threats may exploit vulnerabilities and how the resulting risk will be reduced.
C. Evaluation of the consequences of a risk—in combination with the likelihood of a risk—is important for the prioritization of risk responses. However, it is not the best choice because it does not sufficiently address how the risk will be addressed.
D. **A list of appropriate information security controls in response to the risk scenarios identified during the risk assessment is one of the primary deliverables of a risk assessment exercise. In this case it is also the best choice because it demonstrates due consideration of the risk as well as suitable controls to address the risk.**

R1-37 Which of the following is the **MAIN** outcome of a business impact analysis (BIA)?

A. Project prioritization
B. Criticality of business processes
C. The root cause of IT risk
D. Third-party vendor risk

B is the correct answer.

Justification:
A. Project prioritization is a core focus of program management with a focus on optimizing resource utilization; it is not the main outcome of a BIA.
B. **A BIA measures the total impact of tangible and intangible assets on business processes. Therefore, the sum of the value and opportunity lost as well as the investment and time required to recover is measured to determine the criticality of business processes.**
C. A root cause analysis is a process of diagnosis to establish origins of events, which can be used to learn from consequences, typically from errors and problems, and is not an outcome of a BIA.
D. Third-party vendor risk should be documented during the BIA process, but it is not a main outcome.

DOMAIN 1—RISK IDENTIFICATION, ASSESSMENT AND EVALUATION

R1-38 The **PRIMARY** focus of managing IT-related business risk is to protect:

A. information.
B. hardware.
C. applications.
D. databases.

A is the correct answer.

Justification:
A. **The primary objective for any enterprise is to protect their mission-critical information based on a risk assessment.**
B. While many enterprises spend large amounts protecting their IT hardware, doing so without assessing the risk to their mission-critical data is not advisable. Hardware may become a focus if it stores, processes or transfers mission-critical data.
C. Applications become a focus only if they process mission-critical data.
D. Databases become a focus only if they store mission-critical data.

R1-39 Which of the following provides the **BEST** view of risk management?

A. An interdisciplinary team
B. A third-party risk assessment service provider
C. The enterprise's IT department
D. The enterprise's internal compliance department

A is the correct answer.

Justification:
A. **Having an interdisciplinary team contribute to risk management ensures that all areas are adequately considered and included in the risk assessment processes to support an enterprisewide view of risk.**
B. Engaging a third party to perform a risk assessment may provide additional expertise to conduct the risk assessment; but without internal knowledge, it will be difficult to assess the adequacy of the risk assessment performed.
C. A risk assessment performed by the enterprise's IT department is unlikely to reflect the view of the entire enterprise.
D. The internal compliance department ensures the implementation of risk responses based on the requirement of management. It generally does not take an active part in implementing risk responses for items that do not have regulatory implications.

R1-40 Who is **MOST** likely responsible for data classification?

A. The data user
B. The data owner
C. The data custodian
D. The system administrator

B is the correct answer.

Justification:
A. The data user is granted access based on a justified business need and after approval from the data owner.
B. **The data owner is responsible for classifying data according to the enterprise's data classification scheme. The classification scheme then defines who is eligible to access the data and what controls are required.**
C. The data custodian is responsible for the safe custody, transport and storage of the data (and implementation of business rules).
D. System administrators are considered data custodians because they ensure the safe custody, transport and storage of the data (and implementation of business rules).

DOMAIN 1—RISK IDENTIFICATION, ASSESSMENT AND EVALUATION

R1-41 Which of the following provides the **MOST** valuable input to incident response efforts?

A. Qualitative analysis of threats
B. The annual loss expectancy (ALE) total
C. A vulnerability assessment
D. Penetration testing

A is the correct answer.

Justification:
A. **Qualitative analysis of threats is an intuitive view of the outcome of various threat sources. Knowing the kinds of incidents that may occur in order of consequence will be of great benefit to incident response efforts.**
B. The ALE total is the total cost associated with each source of risk and its probability of occurrence. This total may be of interest when preparing the budget, but cannot be directly linked to incident response efforts.
C. A vulnerability assessment is used to determine how easily security can be breached. This provides data about risk.
D. Penetration testing is used to provide tangible evidence that existing vulnerabilities can be exploited and the degree of difficulty to exploit them.

R1-42 Which of the following approaches to corporate policy **BEST** supports an enterprise's expansion to other regions, where different local laws apply?

A. A global policy that does not contain content that might be disputed at a local level
B. A global policy that is locally amended to comply with local laws
C. A global policy that complies with law at corporate headquarters and that all employees must follow
D. Local policies to accommodate laws within each region

B is the correct answer.

Justification:
A. Having one global policy that attempts to address local requirements for all locales is nearly impossible and generally cost prohibitive.
B. **This choice is correct because it is the only way to minimize the effort and also be in line with local laws.**
C. Developing a policy that exclusively takes into account the laws of the corporate headquarters and that does not take into account local laws and regulations will expose the enterprise to various legal actions as well as cause political and reputational loss issues.
D. Local policies for each region, otherwise known as a decentralized approach, requires the enterprise to separately maintain and test a set of documentation/processes for each region, which is extremely expensive. It also does not give an enterprise the opportunity to leverage a set of common practices as is possible by having a global policy that is amended locally.

DOMAIN 1—RISK IDENTIFICATION, ASSESSMENT AND EVALUATION

R1-43 Which of the following is the **BEST** indicator that incident response training is effective?

A. Decreased reporting of security incidents to the incident response team
B. Increased reporting of security incidents to the incident response team
C. Decreased number of password resets
D. Increased number of identified system vulnerabilities

B is the correct answer.

Justification:
A. Decreased reporting is a sign that users are unaware of what constitutes a security incident.
B. **Increased reporting of incidents is a good indicator of user awareness, but increased reporting of valid incidents is the best indicator because it is a sign that users are aware of the security rules and know how to report incidents. It is the responsibility of the IT function to assess the information provided, identify false-positives, educate end users, and respond to potential problems.**
C. A decrease in the number of password resets is not an indicator of security awareness training.
D. An increase in the number of system vulnerabilities is not an indicator of security awareness training.

R1-44 Which of the following factors will have the **GREATEST** impact on the type of information security governance model that an enterprise adopts?

A. The number of employees
B. The enterprise's budget
C. The organizational structure
D. The type of technology that the enterprise uses

C is the correct answer.

Justification:
A. The number of employees has less impact on information security governance models because well-defined processes provide the proper governance.
B. Organizational budget does not have a major impact on an enterprise's choice of an information security governance model.
C. **Information security governance models are highly dependent on the overall organizational structure.**
D. The type of technology that the enterprise uses has less impact on information security governance models because well-defined processes provide the proper governance.

R1-45 An enterprise has learned of a security breach at another entity that utilizes similar technology. The **MOST** important action a risk practitioner should take is to:

A. assess the likelihood of the incident occurring at the risk practitioner's enterprise.
B. discontinue the use of the vulnerable technology.
C. report to senior management that the enterprise is not affected.
D. remind staff that no similar security breaches have taken place.

A is the correct answer.

Justification:
A. **The risk practitioner should first assess the likelihood of a similar incident occurring at his/her enterprise, based on available information.**
B. Discontinuing the use of the vulnerable technology is not necessarily required; furthermore, the technology is likely to be needed to support the enterprise.
C. Reporting to senior management that the enterprise is not affected is premature until the risk practitioner can first assess the impact of the incident.
D. Until research has been conducted, it is not certain that no similar security breaches have taken place.

DOMAIN 1—RISK IDENTIFICATION, ASSESSMENT AND EVALUATION

R1-46 Which of the following is the **GREATEST** benefit of a risk-aware culture?

A. Issues are escalated when suspicious activity is noticed.
B. Controls are double-checked to anticipate any issues.
C. Individuals communicate with peers for knowledge sharing.
D. Employees are self-motivated to learn about costs and benefits.

A is the correct answer.

Justification:
A. **Management will benefit most from an escalation process because they can become aware of risk or incidents in a timely manner. In addition, escalation posture among employees will best be trained through a program such as awareness.**
B. Double-checking controls is a thorough business practice. It is a basic business stance so the benefit on the management side may be limited.
C. Knowledge sharing is an important theme and this idea should be disseminated through an awareness program. However, the benefit on the risk management side may be indirect.
D. Giving employees a mind-set to learn is desirable. However, cost and benefit knowledge enrichment may not be the primary objective that management expects from awareness efforts.

R1-47 The **MAIN** objective of IT risk management is to:

A. prevent loss of IT assets.
B. provide timely management reports.
C. ensure regulatory compliance.
D. enable risk-aware business decisions.

D is the correct answer.

Justification:
A. Protection of IT assets is a subset goal targeted in an IT risk management program.
B. It is true that an adequate IT risk management program adds value to management reports–an example is presentation of a measurable return of IT investment. However, timeliness in reporting is a separate issue to be discussed apart from IT risk management.
C. Meeting regulatory compliance requirements is a part of the objectives to be achieved in an IT risk management framework.
D. **IT risk management should be conducted as part of enterprise risk management (ERM), the ultimate objective of which is to enable risk-aware business decisions.**

DOMAIN 1—RISK IDENTIFICATION, ASSESSMENT AND EVALUATION

R1-48 Which of the following is the **MOST** desirable strategy when developing risk mitigation options associated with the unavailability of IT services due to a natural disaster?

A. Assume the worst-case incident scenarios.
B. Target low-cost locations for alternate sites.
C. Develop awareness focused on natural disasters.
D. Enact multiple tiers of authority delegation.

A is the correct answer.

Justification:
A. **To be prepared for a natural disaster, it is appropriate to assume the worst-case scenario; otherwise, the resulting impact may exceed the enterprise's ability to recover.**
B. Setting up a low-cost location for an alternate site may not always be a good strategy against natural disasters. Adequate investment should be made based on an impact analysis.
C. An awareness training program is a key factor for business continuity. However, its effectiveness may be limited.
D. Delegation of authority will work somewhat in case of emergency. However, this may be a situational decision in the event of natural disaster.

R1-49 Which of the following is the **BEST** risk identification technique for an enterprise that allows employees to identify risk anonymously?

A. The Delphi technique
B. Isolated pilot groups
C. A strengths, weaknesses, opportunities and threats (SWOT) analysis
D. A root cause analysis

A is the correct answer.

Justification:
A. **With the Delphi technique, polling or information gathering is done either anonymously or privately between the interviewer and interviewee.**
B. With isolated pilot groups, participants generally do not anonymously identify risk.
C. With a SWOT analysis, participants generally do not anonymously identify risk.
D. With a root cause analysis, participants generally do not anonymously identify risk.

R1-50 Which of the following is the **GREATEST** challenge of performing a quantitative risk analysis?

A. Obtaining accurate figures on the impact of a realized threat
B. Obtaining accurate figures on the value of assets
C. Calculating the annualized loss expectancy (ALE) of a specific threat
D. Obtaining accurate figures on the frequency of specific threats

D is the correct answer.

Justification:
A. The impact of a threat can be determined based on the type of threat that occurs.
B. The value of an asset should be fairly easy to ascertain.
C. ALE will not be difficult to calculate if you know the correct frequency of the threat occurring.
D. **It can be challenging to obtain an accurate figure on the frequency of a threat occurring.**

DOMAIN 1—RISK IDENTIFICATION, ASSESSMENT AND EVALUATION

R1-51 Which of the following is the **PRIMARY** factor when deciding between conducting a quantitative or qualitative risk assessment?

A. The corporate culture
B. The amount of time available
C. The availability of data
D. The cost involved with risk assessment

C is the correct answer.

Justification:
A. Management will make decisions based on the risk assessment provided. If management makes decisions based only on financial values, then a quantitative risk analysis is appropriate. If the decision will be based on non-numerical values regarding conceptual elements, then a qualitative analysis is appropriate.
B. The amount of time available may be a factor in deciding between a quantitative and qualitative analysis, but it is not the primary factor.
C. The availability of data is the primary factor in deciding between a quantitative and qualitative risk analysis.
D. The cost involved with a risk assessment may be a factor in deciding between a quantitative and qualitative analysis, but it is not the primary factor.

R1-52 Who **MUST** give the final sign-off on the IT risk management plan?

A. IT auditors performing the risk assessment
B. Business process owners
C. Senior management
D. IT security administrators

C is the correct answer.

Justification:
A. IT auditors performing the risk assessment may be involved in creating the risk management plan, but they do not have the authority to give the final sign-off.
B. Business process owners may be involved in creating the risk management plan, but they do not have the authority to give the final sign-off.
C. By understanding the system and subsystem, senior management have knowledge of the performance metrics and indicators used to measure the system and subsystems, of how the policies and standards are applied within the system and subsystems, and an understanding of the risk and potential impacts associated with recent audit findings and recommendations.
D. IT security administrators may be involved in creating the risk management plan, but they do not have the authority to give the final sign-off.

DOMAIN 1—RISK IDENTIFICATION, ASSESSMENT AND EVALUATION

R1-53 Which of the following is the **PRIMARY** reason that a risk practitioner determines the security boundary prior to conducting a risk assessment?

A. To determine which laws and regulations apply
B. To determine the scope of the risk assessment
C. To determine the business owner(s) of the system
D. To decide between conducting a quantitative or qualitative analysis

B is the correct answer.

Justification:
A. The risk assessment itself will take into account the laws and regulations that apply.
B. The primary reason for determining the security boundary is to establish what systems and components are included in the risk assessment.
C. Determining the business owners is a reason for determining the security boundary, but is secondary to determining the scope.
D. The security boundary should not be a primary factor in deciding between conducting a quantitative or qualitative risk analysis.

R1-54 The **MOST** effective method to conduct a risk assessment on an internal system in an organization is to start by understanding the:

A. performance metrics and indicators.
B. policies and standards.
C. recent audit findings and recommendations.
D. system and its subsystems.

D is the correct answer.

Justification:
A. The person performing the risk assessment should already understand the performance metrics and indicators.
B. The person performing the risk assessment should already understand the policies and standards of the organization.
C. Recent audit findings and recommendations could be useful, but are not as important as understanding the system.
D. To conduct a proper risk assessment, the risk practitioner must understand the system, subsystems and how they work. This provides knowledge of how policies and standards are applied within the system and subsystems, an understanding of process-specific risk, existing interdependencies, and performance indicators.

DOMAIN 1—RISK IDENTIFICATION, ASSESSMENT AND EVALUATION

R1-55 Which of the following **BEST** describes the information needed for each risk on a risk register?

A. Various risk scenarios with their date, description, impact, probability, risk score, mitigation action and owner
B. Various risk scenarios with their date, description, risk score, cost to remediate, communication plan, and owner
C. Various risk scenarios with their date, description, impact, cost to remediate, and owner
D. Various activities leading to risk management planning

A is the correct answer.

Justification:
A. **This choice is the best statement because it contains the necessary elements of the risk register that are needed to make informed decisions.**
B. This choice contains some elements of a risk register, but misses out on some important and key elements of a risk register (impact, probability, mitigation action) that are needed to make informed decisions and this choice lists some items that should not be included in the register (communication plan).
C. This choice misses some important and key elements of a risk register (probability, risk score, mitigation action) needed to make informed decisions.
D. A risk register is a result of risk management planning, not the other way around.

R1-56 Which of the following **BEST** describes the risk-related roles and responsibilities of an organizational business unit (BU)? The BU management team:

A. owns the mitigation plan for the risk belonging to their BU, while board members are responsible for identifying and assessing risk as well as reporting on that risk to the appropriate support functions.
B. owns the risk and is responsible for identifying, assessing and mitigating risk as well as reporting on that risk to the appropriate support functions and the board of directors.
C. carries out the respective risk-related responsibilities, but ultimate accountability for the day-to-day work of risk management and goal achievement belongs to the board members.
D. is ultimately accountable for the day-to-day work of risk management and goal achievement, and board members own the risk.

B is the correct answer.

Justification:
A. This choice is incorrect because the BU management team owns both the risk management activities (identifying, assessing and reporting the mitigation plan for the risk belonging to their BU) and the reporting activities. The board members do not perform the risk identification, assessment and risk reporting functions.
B. **This choice is the best statement because it assigns a senior management level owner to the risk and its resulting actions. Risk owners have the responsibility of identifying, measuring, monitoring, controlling and reporting on risk to executive management as established by the corporate risk framework.**
C. This choice is incorrect because the ultimate accountability for the day-to-day work also belongs to the BU.
D. This choice is incorrect in the sense that it is reversed. The board members do not own the BU risk; the BU leader owns it, and along with the BU management team is accountable for the remediation efforts.

R1-57 The **PRIMARY** advantage of creating and maintaining a risk register is to:

A. ensure that an inventory of potential risk is maintained.
B. ensure that all assets have a low residual risk.
C. define the risk assessment methodology.
D. study a variety of risk and redefine the threat landscape.

A is the correct answer.

Justification:
A. **Once important assets and the risk that may impact these assets are identified, the risk register is used as an inventory of that risk. The risk register can help enterprises accelerate their risk decision making and establish accountability for specific risk.**
B. Simply creating and maintaining a risk register without actual controls deployment does not lower residual risk.
C. The risk assessment methodology includes creation of a risk register as one step in the process; however, this specific step does not define the overall process.
D. The purpose of a risk register is not to enable the study of risk and threats; these are identified during the risk assessment, which provides meaningful input into the risk register.

R1-58 The **GREATEST** advantage of performing a business impact analysis (BIA) is that it:

A. does not have to be updated because the impact will not change.
B. promotes continuity awareness in the enterprise.
C. can be performed using only qualitative estimates.
D. eliminates the need to perform a risk analysis.

B is the correct answer.

Justification:
A. A BIA will need to be updated periodically because systems do change and new systems are added.
B. **A BIA raises the level of awareness for business continuity within the enterprise.**
C. A BIA should utilize both qualitative and quantitative estimates.
D. A BIA does not eliminate the need to perform a risk analysis.

DOMAIN 1—RISK IDENTIFICATION, ASSESSMENT AND EVALUATION

R1-59 When using a formal approach to respond to a security-related incident, which of the following provides the **GREATEST** benefit from a legal perspective?

A. Proving adherence to statutory audit requirements
B. Proving adherence to corporate data protection requirements
C. Demonstrating due care
D. Working with law enforcement agencies

C is the correct answer.

Justification:
A. There are several reasons why adherence to statutory audit requirements will not provide the greatest benefit:
   - Statutory audit requirements do not dictate information security or information technology policies, which define the controls in place to protect the enterprise from security incidents.
   - The response to audit requirements is often a reactive approach.
B. One must implement security to ensure data protection. While this is a good option, it will not offer the greatest benefit because enterprises can have data protection requirements (i.e., policies, etc.) but fail to implement the information security actions to ensure data protection.
C. In the field of information security, the following statements are useful: *"Due care are steps that are taken to show that a company has taken responsibility for the activities that take place within the corporation and has taken the necessary steps to help protect the company, its resources, and employees."* And, *"continual activities that make sure the protection mechanisms are continually maintained and operational."* (Source: Harris, Shon; *All-in-one CISSP Certification Exam Guide, 2nd Edition*, McGraw-Hill/Osborne, USA, 2003.) Stockholders, customers, business partners and governments have the expectation that corporate officers will run the business in accordance with accepted business practices and in compliance with laws and other regulatory requirements. So while no entity can protect themselves completely from security incidents, in case of legal action, by demonstrating due care, these entities can make a case that they are actually doing things to monitor and maintain the protection mechanisms and that these activities are ongoing.
D. Working with law enforcement agencies often occurs after a breach or security incident happens, i.e., a reactive approach. While it is a commendable action, it still raises the question, often in the court of law, of whether the enterprise did everything it could to prevent the breach or incident.

R1-60 Which of the following processes is **CRITICAL** for deciding prioritization of actions in a business continuity plan (BCP)?

A. Risk assessment
B. Vulnerability assessment
C. A business impact analysis (BIA)
D. Business process mapping

C is the correct answer.

Justification:
A. Risk assessment provides information on the likelihood of occurrence of security incidents and assists in the selection of countermeasures, but not in the prioritization of actions.
B. A vulnerability assessment provides information regarding the security weaknesses of the system and supporting the risk analysis process.
C. The BIA is the most critical process for deciding which part of the information system/business process should be given prioritization in case of a security incident.
D. Business process mapping does not help in making a decision, but in implementing a decision.

DOMAIN 1—RISK IDENTIFICATION, ASSESSMENT AND EVALUATION

R1-61 Which of the following is **MOST** effective in assessing business risk?

A. A use case analysis
B. A business case analysis
C. Risk scenarios
D. A risk plan

C is the correct answer.

Justification:
A. A use case analysis is the most common technique used to identify the requirements of a system and the information used to define processes.
B. Business cases are generally a part of the project charter and help define the purpose/reason for the project.
C. Risk scenarios are the most effective technique in assessing business risk.
D. A risk plan is the output document from the assessment of risk.

R1-62 The board of directors of a one-year-old start-up company has asked their chief information officer (CIO) to create all of the enterprise's IT policies and procedures. Which of the following should the CIO create **FIRST**?

A. The strategic IT plan
B. The data classification scheme
C. The information architecture document
D. The technology infrastructure plan

A is the correct answer.

Justification:
A. The strategic IT plan is the first policy to be created when setting up an enterprise's governance model.
B. The strategic IT plan is created before the data classification scheme is developed. The data classification scheme is a method for classifying data by factors such as criticality, sensitivity and ownership.
C. The strategic IT plan is created before the information architecture is defined. The information architecture is one component of the IT architecture (together with applications and technology). The IT architecture is a description of the fundamental underlying design of the IT components of the business, the relationships among them, and the manner in which they support the organization's objectives.
D. The strategic IT plan is created before the technology infrastructure plan is developed. The technology infrastructure plan maps out the technology, human resources and facilities that enable the current and future processing and use of applications.

DOMAIN 2—RISK RESPONSE (17%)

R2-1 Because of its importance to the business, an enterprise wants to quickly implement a technical solution that deviates from the company's policies. The risk practitioner should:

A. recommend against implementation because it violates the company's policies.
B. recommend revision of the current policy.
C. conduct a risk assessment and allow or disallow based on the outcome.
D. recommend a risk assessment and subsequent implementation only if residual risk is accepted.

D is the correct answer.

Justification:
A. Every business decision is driven by cost and benefit considerations. A risk practitioner's contribution to the process is most likely a risk assessment, identifying both the risk and opportunities related to the proposed solution.
B. A recommendation to revise the current policy should not be triggered by a single request.
C. While a risk practitioner may conduct a risk assessment to enable a risk-aware business decision, it is management who will make the final decision.
D. A risk assessment should be conducted to clarify the risk whenever the company's policies cannot be followed. The solutions should only be implemented if the related risk is formally accepted by the business.

R2-2 When proposing the implementation of a specific risk mitigation activity, a risk practitioner **PRIMARILY** utilizes a:

A. technical evaluation report.
B. business case.
C. vulnerability assessment report.
D. budgetary requirements.

B is the correct answer.

Justification:
A. A technical evaluation report is supplemental to a business case.
B. A manager needs to base the proposed risk response on a risk evaluation, the business need (new product, changes in process, compliance need, etc.) and the requirements of the enterprise (new technology, cost, etc.). The manager must look at the costs of the various controls and compare them against the benefit that the organization will receive from the risk response. The manager needs to have knowledge of business case development to illustrate the costs and benefits of the risk response.
C. A vulnerability assessment report is supplemental to a business case.
D. Budgetary requirements are an input into a business case.

DOMAIN 2—RISK RESPONSE

R2-3 A new regulation for safeguarding information processed by a specific type of transaction has come to the attention of an IT manager. The manager should **FIRST**:

A. meet with stakeholders to decide how to comply.
B. analyze the key risk in the compliance process.
C. update the existing security/privacy policy.
D. assess whether existing controls meet the regulation.

D is the correct answer.

Justification:
A. Meeting with stakeholders is a subsequent action to understanding the impact and requirements and performing a gap assessment.
B. Analyzing the key risk in the compliance process is a subsequent action to understanding the impact and requirements and performing a gap assessment.
C. Updating the existing security/privacy policy is a subsequent action to the understanding the impact and requirements and performing a gap assessment.
D. **The first step is to understand the impact and requirements of the new regulation, which includes assessing how the enterprise will comply with the regulation and to what extent the existing control structure supports the compliance process. The risk practitioner should then assess any existing gaps.**

R2-4 Risk management programs are designed to reduce risk to:

A. the point at which the benefit exceeds the expense.
B. a level that is too small to be measurable.
C. a rate of return that equals the current cost of capital.
D. a level that the enterprise is willing to accept.

D is the correct answer.

Justification:
A. Depending on the risk preference of an enterprise, it may or may not choose to pursue risk mitigation to the point at which the benefit equals or exceeds the expense.
B. Reducing risk to a level too small to measure is not practical and is often cost-prohibitive.
C. Tying risk to a specific rate of return ignores the qualitative aspects of risk that must also be considered.
D. **Risk should be reduced to a level that an organization is willing to accept.**

R2-5 Whether a risk has been reduced to an acceptable level should be determined by:

A. IS requirements.
B. information security requirements.
C. international standards.
D. organizational requirements.

D is the correct answer.

Justification:
A. IS requirements should not make the ultimate determination.
B. Information security requirements should not make the ultimate determination.
C. Because each enterprise is unique, international standards of best practice do not represent the best solution.
D. **Organizational requirements should determine when a risk has been reduced to an acceptable level.**

R2-6 Risk assessment techniques should be used by a risk practitioner to:

A. maximize the return on investment (ROI).
B. provide documentation for auditors and regulators.
C. justify the selection of risk mitigation strategies.
D. quantify the risk that would otherwise be subjective.

C is the correct answer.

Justification:
A. Maximizing ROI may be a key objective for implementing risk responses, but is not part of the risk assessment process.
B. A risk assessment does not focus on auditors or regulators as primary recipients of the risk assessment documentation. However, risk assessment results may provide input into the audit process.
C. **A risk practitioner should use risk assessment techniques to justify and implement a risk mitigation strategy as efficiently as possible.**
D. Risk assessment is generally high-level, whereas risk analysis can be either quantitative or qualitative, based on the needs of the organization.

R2-7 Which of the following is the **BEST** method to ensure the overall effectiveness of a risk management program?

A. Assignment of risk within the enterprise
B. Comparison of the program results with industry standards
C. Participation by applicable members of the enterprise
D. User assessment of changes in risk

C is the correct answer.

Justification:
A. Assignment of risk within the enterprise is important to ensure that risk owners are clearly defined and aware of their responsibilities.
B. Comparison of the program results with industry standards may result in valuable feedback—similar to benchmarking—but is not as important as stakeholder participation.
C. **Effective risk management requires participation, support and acceptance by all applicable members of the enterprise, beginning with the executive levels. Personnel must understand their responsibilities and be trained on how to fulfill their roles.**
D. User assessment of changes is a subjective method of assessing risk and not part of a mature risk management program.

DOMAIN 2—RISK RESPONSE

R2-8 Which of the following is the **MOST** effective way to treat a risk such as a natural disaster that has a low probability and a high impact level?

A. Eliminate the risk.
B. Accept the risk.
C. Transfer the risk.
D. Implement countermeasures.

C is the correct answer.

Justification:
A. Eliminating the risk may not be possible.
B. Accepting the risk leaves the enterprise vulnerable to a catastrophic disaster that could cripple or ruin the organization.
C. **Typically, when the probability of an incident is low, but the impact is high, risk is transferred to insurance companies. Examples include hurricanes, tornados and earthquakes. While an enterprise cannot technically transfer risk, transferring risk describes a risk response in which an enterprise indemnifies against the impact of the realized risk.**
D. Implementing countermeasures may not be the most cost-effective approach to security management. It would be more cost-effective to pay recurring insurance costs than to be affected by a disaster from which the enterprise cannot financially recover.

R2-9 A risk response report includes recommendations for:

A. acceptance.
B. assessment.
C. evaluation.
D. quantification.

A is the correct answer.

Justification:
A. **Acceptance of a risk is an alternative to be considered in the risk response process.**
B. The risk assessment process is completed prior to determining appropriate risk responses.
C. Risk evaluation is part of the risk assessment process that is completed prior to determining appropriate risk responses.
D. Risk quantification is achieved during risk analysis; it is an input into the risk response process and occurs prior to determining appropriate risk responses.

R2-10 Which of the following is minimized when acceptable risk is achieved?

A. Transferred risk
B. Control risk
C. Residual risk
D. Inherent risk

C is the correct answer.

Justification:
A. Transferred risk is risk that has been shared with a third party, such as an insurance provider; it may not necessarily be equal to the minimal amount of residual risk.
B. Control risk is the risk that controls may not meet the control objective.
C. **After putting into place an effective risk management program, the remaining risk is called residual risk. Acceptable risk is achieved when residual risk is minimized.**
D. Inherent risk is a risk that is part of an activity; it cannot be minimized, only avoided by not engaging in the activity itself.

DOMAIN 2—RISK RESPONSE

R2-11 During a risk management exercise, an analysis was conducted on the identified risk and mitigations were identified. Which choice **BEST** reflects residual risk?

A. Risk left after the implementation of new or enhanced controls
B. Risk mitigated as a result of the implementation of new or enhanced controls
C. Risk identified prior to implementation of new or enhanced controls
D. Risk classified as high after the implementation of new or enhanced controls

A is the correct answer.

Justification:
A. **The classic definition of residual risk is any risk left after appropriate controls have been implemented to mitigate the target risk.**
B. Residual risk is the risk that remains and is not mitigated.
C. Risk is not identified prior, but after, the implementation of controls.
D. Residual risk can be rated at any level, not just high risk.

R2-12 A global financial institution has decided not to take any further action on a denial of service (DoS) risk found by the risk assessment team. The **MOST** likely reason for making this decision is that:

A. the needed countermeasure is too complicated to deploy.
B. there are sufficient safeguards in place to prevent this risk from happening.
C. the likelihood of the risk occurring is unknown.
D. the cost of countermeasure outweighs the value of the asset and potential loss.

D is the correct answer.

Justification:
A. While countermeasures can be too complicated to deploy, this is not the most compelling reason.
B. Any safeguards placed to prevent the risk need to match the risk impact.
C. It is likely that a global financial institution may be exposed to such DoS attacks, but the frequency cannot be predicted.
D. **An enterprise may decide to accept a specific risk because the protection would cost more than the potential loss.**

R2-13 Which of the following is **MOST** relevant to include in a cost-benefit analysis of a two-factor authentication system?

A. The approved budget of the project
B. The frequency of incidents
C. The annual loss expectancy (ALE) of incidents
D. The total cost of ownership (TCO)

D is the correct answer.

Justification:
A. The approved budget of the project may have no bearing on what the project may actually cost.
B. The frequency and ALE of incidents can help measure the benefit, but have more of an indirect relationship because not all incidents may be mitigated by implementing a two-factor authentication system.
C. The frequency and ALE of incidents can help measure the benefit, but have more of an indirect relationship because not all incidents may be mitigated by implementing a two-factor authentication system.
D. **TCO is the most relevant piece of information to be included in the cost-benefit analysis because it establishes a cost baseline that must be considered for the full life cycle of the control.**

DOMAIN 2—RISK RESPONSE

R2-14 After the completion of a risk assessment, it is determined that the cost to mitigate the risk is much greater than the benefit to be derived. A risk practitioner should recommend to business management that the risk be:

A. treated.
B. terminated.
C. accepted.
D. transferred.

C is the correct answer.

Justification:
A. Treating the risk in the described scenario incurs a cost that is greater than the benefit to be derived; this is not the best option.
B. Risk termination is not a risk management term; while risk can be avoided, it can generally not be terminated.
C. **When the cost of control is more than the cost of the potential impact, the risk should be accepted.**
D. Transferring risk is of limited benefit if the cost of the risk response is more than the cost of the potential likelihood and impact of the risk itself.

R2-15 Which of the following choices will **BEST** protect the enterprise from financial risk?

A. Insuring against the risk
B. Updating the IT risk registry
C. Improving staff training in the risk area
D. Outsourcing the process to a third party

A is the correct answer.

Justification:
A. **An insurance policy can compensate the enterprise up to 100 percent.**
B. Updating the risk registry (with lower values for impact and probability) will not change the risk, only management's perception of it.
C. Staff capacity to detect or mitigate the risk may potentially reduce the financial impact, but never cover it fully the way insurance can.
D. Outsourcing the process containing the risk does not necessarily remove or change the risk.

R2-16 A **PRIMARY** reason for initiating a policy exception process is when:

A. the risk is justified by the benefit.
B. policy compliance is difficult to enforce.
C. operations are too busy to comply.
D. users may initially be inconvenienced.

A is the correct answer.

Justification:
A. **Exceptions to policy are warranted in circumstances in which the benefits outweigh the costs of policy compliance; however, the enterprise needs to asses both the tangible and intangible risk and assess those against the existing risk.**
B. Difficult enforcement is not justification for policy exceptions.
C. Busy operations are not justification for policy exceptions.
D. User inconvenience is not a reason to automatically grant exception to a policy.

DOMAIN 2—RISK RESPONSE

R2-17 The preparation of a risk register begins in which risk management process?

A. Risk response planning
B. Risk monitoring and control
C. Risk management planning
D. Risk identification

D is the correct answer.

Justification:
A. In the risk response planning process, appropriate responses are chosen, agreed-on and included in the risk register.
B. Risk monitoring and control often requires identification of new risk and reassessment of risk. Outcomes of risk reassessments, risk audits and periodic risk reviews trigger updates to the risk register.
C. Risk management planning describes how risk management will be structured and performed.
D. **The risk register details all identified risk, including description, category, cause, probability of occurring, impact(s) on objectives, proposed responses, owners and current status. The primary outputs from risk identification are the initial entries into the risk register.**

R2-18 A risk practitioner receives a message late at night that critical IT equipment will be delivered several days late due to flooding. Fortunately, a reciprocal agreement exists with another company for a replacement until the equipment arrives. This is an example of risk:

A. transfer.
B. avoidance.
C. acceptance.
D. mitigation.

D is the correct answer.

Justification:
A. Risk transfer is not the correct answer because the described risk is not transferred using insurance or another risk transfer strategy.
B. Arranging for a standby is a risk mitigation strategy, not a risk avoidance strategy.
C. The risk is not accepted; if it were accepted, the enterprise would, for example, continue operating without the expected IT equipment until it was delivered.
D. **Risk mitigation attempts to reduce the impact when a risk event occurs. Making plans such as a reciprocal arrangement with another company reduces the consequence of the risk event.**

R2-19 Which of the following would **BEST** help an enterprise select an appropriate risk response?

A. The degree of change in the risk environment
B. An analysis of risk that can be transferred were it not eliminated
C. The likelihood and impact of various risk scenarios
D. An analysis of control costs and benefits

D is the correct answer.

Justification:
A. The degree of change in the risk environment will not provide information of actual controls and benefits to make the decision.
B. Risk can never be eliminated and even analysis of what risk can be transferred will be inadequate for a complete risk response.
C. Likelihood and impact help establish the amount or level of risk.
D. **An analysis of costs and benefits for controls helps an enterprise understand if it can mitigate the risk to an acceptable level.**

DOMAIN 2—RISK RESPONSE

R2-20 In the risk management process, a cost-benefit analysis is **MAINLY** performed:

A. as part of an initial risk assessment.
B. as part of risk response planning.
C. during an information asset valuation.
D. when insurance is calculated for risk transfer.

B is the correct answer.

Justification:
A. A cost-benefit analysis is not only performed once, but every time controls need to be selected to address new or changing risk.
B. **In risk response, a range of controls will be identified that can mitigate the risk; however, a cost-benefit analysis in this process will help identify the right controls that will address the risk at acceptable levels within the budget.**
C. In information asset valuation, business owners determine the value based on business importance and there is no cost-benefit involved.
D. Calculating insurance for the purpose of transferring risk is not the stage where a cost-benefit analysis is performed.

R2-21 Which of the following leads to the **BEST** optimal return on security investment?

A. Deploying maximum security protection across all of the information assets
B. Focusing on the most important information assets and then determining their protection
C. Deploying minimum protection across all the information assets
D. Investing only after a major security incident is reported to justify investment

B is the correct answer.

Justification:
A. Deploying maximum controls across all information assets will overprotect some of the less critical information assets; therefore, investment will not be optimized.
B. **To optimize return on security investment, the primary focus should be identifying the important information assets and protecting them appropriately to optimize investment (i.e., important information assets get more protection than less important or critical assets).**
C. Deploying minimum protection across all the information assets will compromise the security of the more critical information assets; therefore, investment will not be optimized.
D. Investing only after a major security event is a reactive approach that may severely compromise business operations—in some cases, to the extent where the business does not survive.

R2-22 In a situation where the cost of anti-malware exceeds the loss expectancy of malware threats, what is the **MOST** viable risk response?

A. Risk elimination
B. Risk acceptance
C. Risk transfer
D. Risk mitigation

B is the correct answer.

Justification:
A. Risk elimination is not a risk response because it is not possible to reduce risk to zero.
B. **When the cost of a risk response (i.e., the implementation of anti-malware) exceeds the loss expectancy, the most viable risk response is risk acceptance.**
C. Transferring risk to a third party is most viable in situations where the potential likelihood is low and the potential impact is high. Transfer of risk—like any risk response—must be based on a cost-benefit analysis. If the cost of the risk exceeds the cost of the expected loss, the most viable risk response is to accept the risk.
D. Risk mitigation is a method to reduce the likelihood and/or impact of risk to an acceptable level. Risk mitigation—like any risk response—must be based on a cost-benefit analysis. If the cost of the risk exceeds the cost of the expected loss, the most viable risk response is to accept the risk.

R2-23 The **PRIMARY** result of a risk management process is:

A. a defined business plan.
B. input for risk-aware decisions.
C. data classification.
D. minimized residual risk.

B is the correct answer.

Justification:
A. Risk management deliverables are not the primary input into the business plan.
B. **Risk management identifies and prioritizes risk and relates the aggregated risk to the enterprise's risk appetite and risk tolerance levels to enable risk-aware decision making.**
C. Establishing classification levels is one of the outputs of the outcome of risk assessment, but is not the primary result.
D. Residual risk is reduced after taking the cost of the risk response and the related benefit into consideration; risk minimization itself is not a primary result of risk management because it may not optimize overall business results.

DOMAIN 2—RISK RESPONSE

R2-24 As part of fire drill testing, designated doors swing open, as planned, to allow employees to leave the building faster. An observer notices that this practice allows unauthorized personnel to enter the premises unnoticed.
The **BEST** way to alter the process is to:

A. stop the designated doors from opening automatically in case of a fire.
B. include the local police force to guard the doors in case of fire.
C. instruct the facilities department to guard the doors and have staff show their badge when exiting the building.
D. assign designated personnel to guard the doors once the alarm sounds.

D is the correct answer.

Justification:
A. Stopping the doors from opening in case of a fire does not effectively support the primary objective of a fire drill, which is to protect human life in case a fire occurs.
B. This choice is not useful because the police have better things to do.
C. Having the facilities department guard the exit doors and monitor staff as they leave the facility does not address the risk of having unauthorized personnel entering the building.
D. **Unless there are designated personnel monitoring each door from the time the alarm sounds, there is no way to prevent unauthorized individuals from entering the building while employees are exiting.**

R2-25 During a quarterly interdepartmental risk assessment, the IT operations center indicates a heavy increase of malware attacks. Which of the following recommendations to the business is **MOST** appropriate?

A. Contract with a new anti-malware software vendor because the current solution seems ineffective.
B. Close down the Internet connection to prevent employees from visiting infected web sites.
C. Make the number of malware attacks part of each employee's performance metrics.
D. Increase employee awareness training, including end-user roles and responsibilities.

D is the correct answer.

Justification:
A. Anti-malware software is always a step behind the malware that exists in the marketplace. This is particularly true for zero-day exploits; because the IT operation center is aware of the attack, the anti-malware in place seems to be effective.
B. Closing down the Internet connections may impair valid business processes and does not provide protection from the variety of channels that malware uses for attack.
C. Making employees responsible for the number of malware attacks that the enterprise is exposed to is an example of incentive misalignment because it punishes employees for something for which they are not responsible.
D. **Employee awareness training will help the enterprise avoid, and more quickly detect, malware attacks, particularly when staff understand the typical symptoms and are knowledgeable about the incident reporting process.**

R2-26 Which of the following is a behavior of risk avoidance?

A. Take no action against the risk.
B. Outsource the related process.
C. Insure against a specific event.
D. Exit the process that gives rise to risk.

D is the correct answer.

Justification:
A. Taking no action is an example of risk acceptance where no action is taken relative to a particular risk, and loss is accepted when/if it occurs. This is different from being ignorant of risk; accepting risk assumes that the risk is known, i.e., an informed decision has been made by management to accept it as such.
B. Outsourcing a process is an example of risk transfer/sharing. It reduces risk frequency or impact by transferring or otherwise sharing a portion of the risk. In both a physical and legal sense this risk transfer does not relieve an enterprise of a risk, but can involve the skills of another party in managing the risk and thus reduce the financial consequence if an adverse event occurs.
C. Insuring against a specific event is an example of risk transfer/sharing. It reduces risk frequency or impact by transferring or otherwise sharing a portion of the risk. In both a physical and legal sense risk transfer does not relieve an enterprise of a risk, but can involve the skills of another party in managing the risk and thus reduce the financial consequence if an adverse event occurs.
D. Avoidance means exiting the activities or conditions that give rise to risk. Risk avoidance applies when no other risk response is adequate. Some IT-related examples of risk avoidance may include relocating a data center away from a region with significant natural hazards or declining to engage in a very large project when the business case shows a notable risk of failure.

R2-27 A chief information security officer (CISO) has recommended several controls such as anti-malware to protect the enterprise's information systems. Which approach to handling risk is the CISO recommending?

A. Risk transference
B. Risk mitigation
C. Risk acceptance
D. Risk avoidance

B is the correct answer.

Justification:
A. Risk transfer involves transferring the risk to another entity such as an insurance company.
B. By implementing controls the company is trying to decrease risk to an acceptable level, thereby mitigating risk.
C. Risk acceptance involves making an educated decision to accept the risk in the system and taking no action.
D. Risk avoidance involves stopping any activity causing the risk.

DOMAIN 2—RISK RESPONSE

R2-28 Which of the following is **MOST** important for determining what security measures to put in place for a critical information system?

A. The number of threats to the system
B. The level of acceptable risk to the enterprise
C. The number of vulnerabilities in the system
D. The existing security budget

B is the correct answer.

Justification:
A. Determining the number of threats to the system is important; however, it alone will not determine the security measures to put in place.
B. **Determining the level of acceptable risk will allow the enterprise to determine the security measures to put in place.**
C. Determining the number of vulnerabilities in the system is important; however, it alone will not determine the security measures to put in place.
D. Determining how much of the budget is available should not determine the security measures to put in place.

R2-29 Obtaining senior management commitment and support for information security investments can **BEST** be accomplished by a business case that:

A. explains the technical risk to the enterprise.
B. includes industry best practices as they relate to information security.
C. details successful attacks against a competitor.
D. ties security risk to organizational business objectives.

D is the correct answer.

Justification:
A. Senior management will not be as interested in technical risk if they are not tied to the impact on business environment and objectives.
B. Industry best practices are important to senior management but, again, senior management will give them the right level of importance when they are presented in terms of key business objectives.
C. Senior management will not be as interested in examples of successful attacks against a competitor if they are not tied to the impact on business environment and objectives.
D. **Senior management seeks to understand the business justification for investing in security. This can best be accomplished by tying security to key business objectives.**

R2-30 Acceptable risk for an enterprise is achieved when:

A. transferred risk is minimized.
B. control risk is minimized.
C. inherent risk is minimized.
D. residual risk is within tolerance levels.

D is the correct answer.

Justification:
A. Risk transfer is the process of assigning risk to another organization, usually through the purchase of an insurance policy or outsourcing the service. In both a physical and legal sense this risk transfer does not relieve an enterprise of a risk, but can involve the skills of another party in managing the risk and thus reduce the financial consequence if an adverse event occurs.
B. Control risk is the risk that a material error exists that would not be prevented or detected on a timely basis by the system of internal controls.
C. Inherent risk is the risk level or exposure without taking into account the actions that management has taken or might take (e.g., implementing controls). Inherent risk cannot be minimized.
D. **Residual risk is the risk that remains after all controls have been applied; therefore, acceptable risk is achieved when residual risk is aligned with the enterprise risk appetite.**

R2-31 A procurement employee notices that new printer models offered by the vendor keep a copy of all printed documents on a built-in hard disk. Considering the risk of unintentionally disclosing confidential data, the employee should:

A. proceed with the order and configure printers to automatically wipe all the data on disks after each print job.
B. notify the security manager to conduct a risk assessment for the new equipment.
C. seek another vendor that offers printers without built-in hard disk drives.
D. procure printers with built-in hard disks and notify staff to wipe hard disks when decommissioning the printer.

B is the correct answer.

Justification:
A. Wiping hard disks after each job is not appropriate without a prior risk assessment because the data may be useful for forensic investigation and may impact performance of the printer.
B. **Risk assessment is the most appropriate answer because it will result in risk mitigation techniques that are appropriate for organizational risk context and appetite.**
C. Focusing solely on the risk and ignoring the opportunity is not a correct approach. A risk associated with nonvolatile storage is not a sufficient reason for changing vendors because this is a general trend with the printers that brings business benefits in addition to risk that needs to be addressed.
D. Notifying staff is not a sufficient control and does not mitigate the risk from printers being serviced by an external party.

DOMAIN 2—RISK RESPONSE

R2-32 Which of the following situations is **BEST** addressed by transferring risk?

A. An antiquated fire suppression system in the computer room
B. The threat of disgruntled employee sabotage
C. The possibility of the loss of a universal serial bus (USB) removable media drive
D. A building located in a 100-year flood plain

D is the correct answer.

Justification:
A. Although an enterprise may hold insurance policies for physical assets such as computer equipment, the most appropriate risk remediation strategy is to update the fire suppression system.
B. This risk is not readily transferrable. Full risk response planning should be performed for all risk that could happen at any time during routine business activities.
C. This risk is not readily transferrable. Removable media policies and procedures should proactively be in place to mitigate the risk of lost or stolen media.
D. **Purchasing an insurance policy transfers the risk of a flood. Risk transfer is the process of assigning risk to another entity, usually through the purchase of an insurance policy or outsourcing the service.**

SCENARIO 1

A scenario is a mini-case study that describes a situation or an organization and requires candidates to answer one or more questions based on the information provided. A scenario can focus on a specific domain or on several domains. The CRISC exam will include scenarios.

QUESTIONS R2-33 THROUGH R2-34 REFER TO THE FOLLOWING INFORMATION:

The chief information officer (CIO) of an enterprise has just received this year's IT security audit report. The report shows numerous open vulnerability findings on both business-critical and nonbusiness-critical information systems. The CIO briefed the chief executive officer (CEO) and board of directors on the findings and expressed his concern on the impact to the enterprise. He was informed that there are not enough funds to mitigate all of the findings from the report.

R2-33 The CIO should respond to the findings identified in the IT security audit report by mitigating:

A. the most critical findings on both the business-critical and nonbusiness-critical systems.
B. all vulnerabilities on business-critical information systems first.
C. the findings that are the least expensive to mitigate first to save funds.
D. the findings that are the most expensive to mitigate first and leave all others until more funds become available.

B is the correct answer.

Justification:
A. Mitigating the critical findings on the nonbusiness-critical systems is a waste of limited funds.
B. **Mitigating vulnerabilities on business-critical information systems should be completed first to ensure that the business can continue to operate.**
C. The expense of the mitigations should be a secondary factor to the value of the information systems.
D. The expense of the mitigations should be a secondary factor to the value of the information systems.

DOMAIN 2—RISK RESPONSE

SEE INFORMATION PRECEDING QUESTION R2-33

R2-34 Assuming that the CIO is unable to address all of the findings, how should the CIO deal with any findings that remain after available funds have been spent?

A. Create a plan of actions and milestones for open vulnerabilities.
B. Shut down the information systems with the open vulnerabilities.
C. Reject the risk on the open vulnerabilities.
D. Implement compensating controls on the systems with open vulnerabilities.

A is the correct answer.

Justification:
A. **Creating a plan of actions and milestones ensures that there is a plan to mitigate the remaining vulnerabilities over time. It will also identify the order in which the vulnerabilities should be mitigated with target dates for mitigation.**
B. Shutting down the system is not the correct vulnerability mitigation strategy. The vulnerability may be on a mission-critical system.
C. Rejecting the risk is not a risk mitigation strategy.
D. Compensating controls will already be placed on the information systems. Additional compensating controls will require funds which have already been depleted.

DOMAIN 3—RISK MONITORING (17%)

R3-1 Which of the following is the **MOST** important reason for conducting periodic risk assessments?

A. Risk assessments are not always precise.
B. Reviewers can optimize and reduce the cost of controls.
C. Periodic risk assessments demonstrate the value of the risk management function to senior management.
D. Business risk is subject to frequent change.

D is the correct answer.

Justification:
A. Although an assessment can never be perfect and invariably contains some errors, this is not the most important reason for periodic reassessment.
B. Optimizing control cost is an insufficient reason.
C. Demonstrating the value of the risk management function to senior management is an insufficient reason.
D. **Risk is constantly changing, so a previously conducted risk assessment may not include measured risk that has been introduced since the last assessment.**

R3-2 Which of the following is **MOST** essential for a risk management program to be effective?

A. New risk detection
B. A sound risk baseline
C. Accurate risk reporting
D. A flexible security budget

A is the correct answer.

Justification:
A. **Without identifying new risk, other procedures will only be useful for a limited period.**
B. A risk baseline is essential for implementing risk management, but new risk detection is the most essential.
C. Accurate risk reporting is essential for implementing risk management, but new risk detection is the most essential.
D. A flexible security budget is not a reality for most enterprises. A limited security budget is a scope limitation that an effective risk management program works with by prioritizing risk responses.

R3-3 A network vulnerability assessment is intended to identify:

A. security design flaws.
B. zero-day vulnerabilities.
C. misconfigurations and missing updates.
D. malicious software and spyware.

C is the correct answer.

Justification:
A. Security design flaws require a deeper level of analysis.
B. Zero-day vulnerabilities, by definition, are not previously known and, therefore, are undetectable.
C. **A network vulnerability assessment intends to identify known vulnerabilities that are based on common misconfigurations and missing updates.**
D. Malicious software and spyware are normally addressed through antivirus and antispyware policies.

DOMAIN 3—RISK MONITORING

R3-4 Previously accepted risk should be:

A. reassessed periodically because the risk can be escalated to an unacceptable level due to revised conditions.
B. removed from the risk log once it is accepted.
C. accepted permanently because management has already spent resources (time and labor) to conclude that the risk level is acceptable.
D. avoided next time because risk avoidance provides the best protection to the enterprise.

A is the correct answer.

Justification:
A. **Accepted risk should be reviewed regularly to ensure that the initial risk acceptance rationale is still valid within the current business context.**
B. Even risk that has been accepted should be monitored for changing conditions that could alter the original decision.
C. The rationale for the initial risk acceptance may no longer be valid due to change(s), and therefore, risk cannot be accepted permanently.
D. Risk is an inherent part of business, and it is impractical and costly to eliminate all risk.

R3-5 After a risk assessment study, a bank with global operations decided to continue doing business in certain regions of the world where identity theft is widespread. To **MOST** effectively deal with the risk, the business should:

A. implement monitoring techniques to detect and react to potential fraud.
B. make the customer liable for losses if the customer fails to follow the bank's advice.
C. increase its customer awareness efforts in those regions.
D. outsource credit card processing to a third party.

A is the correct answer.

Justification:
A. **Implementing monitoring techniques that will detect and deal with potential fraud cases is the most effective way to deal with this risk.**
B. While making the customer liable for losses is a possible approach, the bank needs to be seen as proactive in managing its risk.
C. While customer awareness will help mitigate the risk, this is not sufficient on its own to control fraud risk.
D. If the bank outsources its processing, the bank still retains liability.

R3-6 Which of the following **BEST** indicates a successful risk management practice?

A. Control risk is tied to business units.
B. Overall risk is quantified.
C. Residual risk is minimized.
D. Inherent risk is eliminated.

C is the correct answer.

Justification:
A. Although tying control risk to business units may improve accountability, it is not as desirable as minimizing residual risk.
B. The fact that overall risk has been quantified does not necessarily indicate the existence of a successful risk management practice.
C. **A successful risk management practice minimizes the residual risk to the enterprise.**
D. It is virtually impossible to eliminate inherent risk.

DOMAIN 3—RISK MONITORING

R3-7 Which of the following **MOST** enables risk-aware business decisions?

A. Robust information security policies
B. An exchange of accurate and timely information
C. Skilled risk management personnel
D. Effective process controls

B is the correct answer.

Justification:
A. Security policies generally focus on protecting the business and do not enable risk-aware business decisions, particularly when the decision affects future business needs.
B. **An exchange of information is a key area for management to be able to make risk-related decisions. Accuracy and timeliness of information are success factors.**
C. Skilled risk management personnel enable risk-aware business decisions, but ideally the flow of information needs to be two-directional, ensuring that risk, loss and vulnerability events are reported and allowing risk management personnel to understand changes in the organization's risk appetite and tolerance.
D. Process controls generally exist for known threats and do not enable risk-based business decisions. Control monitoring, however, involves the dissemination of control information to enable a timely risk response (business decision).

R3-8 Which of the following should be of **MOST** concern to a risk practitioner?

A. Failure to notify the public of an intrusion
B. Failure to notify the police of an attempted intrusion
C. Failure to internally report a successful attack
D. Failure to examine access rights periodically

C is the correct answer.

Justification:
A. Reporting to the public is not a requirement and is dependent on the enterprise's desire, or lack thereof, to make the intrusion known.
B. It is highly unlikely that an attempted intrusion requires notification of the police. Moreover, attempted intrusions are not as significant to the risk practitioner as activities related to successful attacks.
C. **Failure to report a successful intrusion is a serious concern to the risk practitioner and could—in some instances—be interpreted as abetting.**
D. Although the lack of a periodic examination of access rights may be a concern, it does not represent as big a concern as the failure to report a successful attack.

R3-9 Which of the following **MOST** likely indicates that a customer data warehouse should remain in-house rather than be outsourced to an offshore operation?

A. The telecommunications costs may be much higher in the first year.
B. Privacy laws may prevent a cross-border flow of information.
C. Time zone differences may impede communications between IT teams.
D. Software development may require more detailed specifications.

B is the correct answer.

Justification:
A. Higher telecommunications costs are more manageable than privacy laws.
B. **Privacy laws prohibiting the cross-border flow of personally identifiable information (PII) may make it impossible to locate a data warehouse containing customer information in another country.**
C. Time zone differences are more manageable than privacy laws.
D. Typically, software development requires more detailed specifications when dealing with offshore operations.

DOMAIN 3—RISK MONITORING

R3-10 Which of the following is the **FIRST** step when developing a risk monitoring program?

A. Developing key indicators to monitor outcomes
B. Gathering baseline data on indicators
C. Analyzing and reporting findings
D. Conducting a capability assessment

D is the correct answer.

Justification:

A. Developing key indicators to monitor outcomes is necessary, but not the first step. There is no use for indicators if there is no information on what these indicators are going to report.
B. Gathering baseline data on indicators is necessary, but not the first step. There is no use for gathering baseline data if the indicators are not defined.
C. Analyzing and reporting findings is necessary, but not the first step. There is no use for analyzing and reporting findings if the baseline is not there.
D. **This step determines the capacity and readiness of the entity to develop a risk management program. This assessment identifies champions, barriers, owners and contributors to this program, including identifying the overall goal of the program. A capability assessment helps determine the enterprise's maturity in its risk management processes and the capacity and readiness of the entity to develop a risk management program. When the enterprise is more mature, more sophisticated responses can be implemented; when the enterprise is rather immature, some basic responses may be a better starting point.**

R3-11 Which of the following reviews will provide the **MOST** insight into an enterprise's risk management capabilities?

A. A capability maturity model (CMM) review
B. A capability comparison with industry standards or regulations
C. A self-assessment of capabilities
D. An internal audit review of capabilities

A is the correct answer.

Justification:
A. **Capability maturity modeling allows an enterprise to understand its level of maturity in its risk capabilities, which is an indicator of operational readiness and effectiveness.**
B. A capability comparison with industry standards or regulations does not provide insights into readiness and effectiveness, but only into the existence or nonexistence of capabilities exclusive of maturity.
C. A self-assessment of capabilities does not provide insights into readiness and effectiveness, but only into the existence or nonexistence of capabilities exclusive of maturity.
D. An internal audit review of capabilities does not provide insights into readiness and effectiveness, but only into the existence or nonexistence of capabilities exclusive of maturity.

DOMAIN 3—RISK MONITORING

R3-12 Which of the following is of **MOST** concern in a review of a virtual private network (VPN) implementation? Computers on the network are located:

A. at the enterprise's remote offices.
B. on the enterprise's internal network.
C. at the backup site.
D. in employees' homes.

D is the correct answer.

Justification:
A. Computers on the network that are at the enterprise's remote offices, perhaps with different IS and security employees who have different ideas about security, may be more risky, but are less risky than an employee's home computer.
B. There should be security policies in place on an enterprise's internal network to detect and halt an outside attack that uses an internal machine as a staging platform.
C. Computers at the backup site are subject to the corporate security policy and, therefore, are not high-risk computers.
D. **In a VPN implementation, there is a risk of allowing high-risk computers onto the enterprise's network. All machines that are allowed onto the virtual network should be subject to the same security policy. Home computers are least often subject to the corporate security policies and, therefore, are high-risk computers. Once a computer is hacked and "owned," any network that trusts that computer is at risk. Implementation and adherence to the corporate security policy is easier when all computers on the network are on the enterprise's campus.**

R3-13 Which type of risk assessment methods involves conducting interviews and using anonymous questionnaires by subject matter experts?

A. Quantitative
B. Probabilistic
C. Monte Carlo
D. Qualitative

D is the correct answer.

Justification:
A. Quantitative risk assessments utilize a mathematical calculation based on security metrics on the asset (system or application).
B. Probabilistic risk assessments use a mathematical model to construct the qualitative risk assessment approach while using the quantitative risk assessment techniques and principles.
C. Monte Carlo simulation combines both qualitative and quantitative assessment approaches and is based on a normal deterministic simulation model.
D. **Qualitative risk assessment methods include using interviewing and the Delphi method, which is the method described in the question.**

DOMAIN 3—RISK MONITORING

R3-14 Which of the following documents **BEST** identifies an enterprise's compliance risk and the corrective actions in progress to meet these regulatory requirements?

A. An internal audit report
B. A risk register
C. An external audit report
D. A risk assessment report

B is the correct answer.

Justification:
A. Audit reports track audit findings and their respective actions, but based on the audit scope, do not necessarily include compliance-oriented findings or their risk. They generally do not include corrective actions in progress.
B. **A risk register provides a report of all current identified risk within an enterprise, including compliance risk, with the status of the corrective actions or exceptions that are associated with them.**
C. External audit reports are generally more reliable than internal audit reports due to the increased independence of the external auditor. Similar to internal audit reports, they do not generally include all relevant compliance risk, but may focus on a single regulatory requirement at a time, such as privacy, the Occupational Safety and Health Administration (OSHA) in the US, the US Sarbanes-Oxley Act of 2002, etc. They generally do not include corrective actions in progress.
D. Risk assessment reports may include compliance risk, but often do not include insights into the corrective actions that are ongoing or planned.

R3-15 Which of the following practices is **MOST** closely associated with risk monitoring?

A. Assessment
B. Mitigation
C. Analysis
D. Reporting

D is the correct answer.

Justification:
A. Risk assessment is associated with risk identification and evaluation, but not with risk monitoring.
B. Risk mitigation is associated with risk response, but not with risk monitoring.
C. Risk analysis is associated with risk identification and evaluation, but not with risk monitoring.
D. **Risk reporting is the only activity listed that is typically associated with risk monitoring.**

R3-16 Which of the following assessments of an enterprise's risk monitoring process will provide the **BEST** information about its alignment with industry-leading practices?

A. A capability assessment by an outside firm
B. A self-assessment of capabilities
C. An independent benchmark of capabilities
D. An internal audit review of capabilities

C is the correct answer.

Justification:
A. A capability assessment by an outside firm does not assess the enterprise against industry peers or competitors and only provides the opinion of the examiner as to what are/are not industry-leading practices.
B. A process capability self-assessment does not assess the enterprise against industry peers or competitors; it provides the opinion of the examiner, which in the case of a self-assessment is not even independent of the process to be reviewed.
C. **An independent benchmark of capabilities allows an enterprise to understand its level of capability compared to other organizations within its industry. This allows the enterprise to identify industry-leading practices and its level of capability associated with these practices.**
D. An internal audit review of capabilities does not assess the enterprise against industry peers or competitors; audits generally measure capabilities against corporate standards, not necessarily against industry-leading practices.

R3-17 Which of the following is the **MOST** prevalent risk in the development of end-user computing (EUC) applications?

A. Increased development and maintenance costs
B. Increased application development time
C. Impaired decision making due to diminished responsiveness to requests for information
D. Applications not subjected to testing and IT general controls

D is the correct answer.

Justification:
A. EUC applications typically result in reduced application development and maintenance costs.
B. EUC applications typically result in a reduced development cycle time.
C. EUC applications normally increase flexibility and responsiveness to management's information requests.
D. **End-user-developed applications may not be subject to an independent outside review by systems analysts and, frequently, are not created in the context of a formal development methodology. These applications may lack appropriate standards, controls, quality assurance procedures and documentation. A risk associated with end-user applications can include backup and recovery procedures that are not applied because operations may not be aware of the application.**

DOMAIN 3—RISK MONITORING

R3-18 As part of an enterprise risk management (ERM) program, a risk practitioner **BEST** leverages the work performed by an internal audit function by having it:

A. design, implement and maintain the ERM process.
B. manage and assess the overall risk awareness.
C. evaluate ongoing changes to organizational risk factors.
D. assist in monitoring, evaluating, examining and reporting on controls.

D is the correct answer.

Justification:
A. The design, implementation and maintenance of the ERM function is the responsibility of management, not of the internal audit function.
B. Overall risk awareness is the responsibility of the risk governance function.
C. Evaluating ongoing changes to the enterprise is not the responsibility of the internal audit function.
D. The internal audit function is responsible for assisting management and the board of directors in monitoring, evaluating, examining and reporting on internal controls, regardless of whether an ERM function has been implemented.

R3-19 Where are key risk indicators (KRIs) **MOST** likely identified when initiating risk management across a range of projects?

A. Risk governance
B. Risk response
C. Risk analysis
D. Risk monitoring

B is the correct answer.

Justification:
A. Risk governance is a systemic approach to decision-making processes associated with risk. From a CRISC perspective, information technology risk is adopted to achieve more effective risk management and to reduce risk exposure and vulnerability by filling gaps in the risk policy. This is not the best answer because it is not a risk management activity, but rather a risk management oversight function.
B. KRIs and risk definition and prioritization are both considered part of the risk response process. After having identified, quantified and prioritized the risk to the enterprise, relevant risk indicators need to be identified to help provide risk owners with meaningful information about a specific risk, or a combination of types of risk.
C. Risk analysis is the process of identifying the types, probability and severity of risk that may occur during a project. Once the identification has taken place, the analysis feeds into the risk response process where one of the tasks is to identify KRIs.
D. Risk monitoring occurs after the risk response process and is ongoing. Assigning ownership to KRIs and defining various levels of KRI thresholds—along with automating the monitoring and notification process—help ensure monitoring of KRIs. KRIs must be identified before risk monitoring is implemented.

DOMAIN 3—RISK MONITORING

R3-20 Risk assessments should be repeated at regular intervals because:

A. omissions in earlier assessments can be addressed.
B. periodic assessments allow various methodologies.
C. business threats are constantly changing.
D. they help raise risk awareness among staff.

C is the correct answer.

Justification:
A. Omissions not found in earlier assessments do not justify regular reassessments.
B. This choice is incorrect because unless the environment changes, risk assessments should be performed using the same methodologies.
C. **As business objectives and methods change, the nature and relevance of threats also change.**
D. This choice is incorrect because there are better ways of raising security awareness than by performing a risk assessment, such as risk awareness training.

R3-21 An operations manager assigns monitoring responsibility of key risk indicators (KRIs) to line staff. Which of the following is **MOST** effective in validating the effort?

A. Reported results should be independently reviewed.
B. Line staff should complete risk management training.
C. The threshold should be determined by risk management.
D. Indicators should have benefits that exceed their costs.

A is the correct answer.

Justification:
A. **Because KRIs are monitored by line staff, there is a chance that staff may alter results to suppress unfavorable results. Additional reliability of monitoring metrics can be achieved by having the results reviewed by an independent party.**
B. It is not mandatory that line staff complete risk management training in order to be engaged in monitoring of KRIs.
C. The threshold should be determined through discussion between risk management and line staff/business managers.
D. It is important that the benefits of KRIs justify their costs; however, this determination does not help verify that the monitoring efforts of KRIs are effective.

R3-22 Which of the following can be expected when a key control is being maintained at an optimal level?

A. The shortest lead time until the control breach comes to the surface
B. Balance between control effectiveness and cost
C. An adequate maturity level of the risk management process
D. An accurate estimation of operational risk amounts

B is the correct answer.

Justification:
A. Even though a key control is in place, it may take time until a breach surfaces if escalation procedures are not adequately set up. Thus, a key control alone does not assure the shortest lead time for a breach to be communicated to management.
B. **Maintaining controls at an optimal level translates into a balance between control cost and derived benefit.**
C. Measurement of the maturity level in risk management may depend on the function of key controls. However, the key control is not the major driver to assess the maturity of risk management.
D. The key control does not directly contribute to the accurate estimation of operational risk amounts. Maintenance of an incident database and the application of statistical method are essential for the estimation of operational risk.

DOMAIN 3—RISK MONITORING

R3-23 The annual expected loss of an asset—the annual loss expectancy (ALE)—is calculated as the:

A. exposure factor (EF) multiplied by the annualized rate of occurrence (ARO).
B. single loss expectancy (SLE) multiplied by the exposure factor (EF).
C. single loss expectancy (SLE) multiplied by the annualized rate of occurrence (ARO).
D. asset value (AV) multiplied by the single loss expectancy (SLE).

C is the correct answer.

Justification:
A. This is not the correct formula to calculate ALE. ALE is calculated by multiplying the single loss expectancy (SLE) by the annualized rate of occurrence (ARO) or the amount of times that the enterprise expects the loss to occur.
B. This is not the correct formula to calculate ALE. ALE is calculated by multiplying the single loss expectancy (SLE) by the annualized rate of occurrence (ARO) or the amount of times that the enterprise expects the loss to occur.
C. **ALE is calculated by is calculated by multiplying the single loss expectancy (SLE) by the annualized rate of occurrence (ARO) or the amount of times that the enterprise expects the loss to occur.**
D. This is not the correct formula to calculate ALE. ALE is calculated by multiplying the single loss expectancy (SLE) by the annualized rate of occurrence (ARO) or the amount of times that the enterprise expects the loss to occur.

R3-24 The **PRIMARY** reason to report significant changes in IT risk to management is to:

A. update the information asset inventory on a periodic basis.
B. update the values of probability and impact for the related risk.
C. reconsider the degree of importance of existing information assets.
D. initiate an appropriate risk response for impacted information assets.

D is the correct answer.

Justification:
A. This choice is not correct because an asset inventory may be updated even when there is no significant risk reported and because this is less important than initiating an appropriate risk response for impacted information assets.
B. This choice is not correct because updating new probability and impact values may happen when the risk assessment is performed or when significant risk is identified and analyzed.
C. This choice is not correct because management staff of relevant functions understand the importance of their assets and do not wait for significant risk to reconsider this fact.
D. **The changes in information risk will impact the business process of a department or multiple departments and the security manager should report this to department heads so that they are able to perform the impact analysis and provide risk acceptance.**

DOMAIN 3—RISK MONITORING

R3-25 Which of the following is **MOST** beneficial to the improvement of an enterprise's risk management process?

A. Key risk indicators (KRIs)
B. External benchmarking
C. The latest risk assessment
D. A maturity model

D is the correct answer.

Justification:
A. KRIs are metrics that help monitor risk over time; they may be used to identify trends, but do not help define the desired state of the enterprise like a maturity model and thus are not the best option.
B. External benchmarking is useful to determine how other, similar enterprises manage risk, but does not help defined the desired state of the enterprise like a maturity model and thus is not the best option.
C. The latest risk assessment will be an input into the risk management process improvement effort, but does not help define the desired state of the enterprise like a maturity model and thus is not the best option.
D. **A maturity model helps identify the status quo as well as the desired state and thus is most helpful when an enterprise desires to improve a business process, such as risk management.**

R3-26 Which of the following is the **PRIMARY** reason for having the risk management process reviewed by independent risk auditors/assessors?

A. To ensure that the risk results are consistent
B. To ensure that the risk factors and risk profile are well defined
C. To correct any mistakes in risk assessment
D. To validate the control weaknesses for management reporting

B is the correct answer.

Justification:
A. Ensuring that risk results are consistent is very important to ensure that risk mitigation/management is effective and that is why risk management results are reviewed by independent risk auditors/assessors, who can be internal or external to the enterprise.
B. **Risk profile and risk factors are defined during the risk assessment process; an independent review helps ensure that the underlying process is effective and helps identify areas for future improvement.**
C. Risk assessment by an independent party is primarily performed to ensure and/or improve the quality of the risk assessment process, not to correct risk assessment mistakes.
D. The primary purpose of independent review is not to validate control weaknesses for management reporting, although it may be an outcome of the process.

DOMAIN 3—RISK MONITORING

R3-27 Which of the following provides the **GREATEST** support to a risk practitioner recommending encryption of corporate laptops and removable media as a risk mitigation measure?

A. Benchmarking with peers
B. Evaluating public reports on encryption algorithm in the public domain
C. Developing a business case
D. Scanning unencrypted systems for vulnerabilities

C is the correct answer.

Justification:
A. Benchmarking with peers does not help because peers will have a different risk environment and culture that cannot directly apply to one's own enterprise.
B. While evaluation of the solution in the public domain is important information, risk practitioners still need to analyze each solution in the context of their enterprise to provide the most valuable recommendation.
C. **A business case has the business reasoning as to why the encryption solutions address the risk and also explains how the risk losses can be reduced.**
D. Conducting a vulnerability assessment of unencrypted systems without proper business justification does not help.

R3-28 A database administrator notices that the externally hosted, web-based corporate address book application requires users to authenticate, but that the traffic between the application and users is not encrypted. The **MOST** appropriate course of action is to:

A. notify the business owner and the security manager of the discovery and propose an addition to the risk register.
B. contact the application administrators and request that they enable encryption of the application's web traffic.
C. alert all staff about the vulnerability and advise them not to log on from public networks.
D. accept that current controls are suitable for nonsensitive business data.

A is the correct answer.

Justification:
A. **The business owner and security manager should be notified and the risk should be documented on the operational or security risk register to enable appropriate risk treatment.**
B. Enabling encryption without further assessment and input from the business owner is inappropriate because the vulnerability may indicate further issues with security that need to be resolved.
C. Alerting all of the staff without discussing the risk with the business owner and having a plan to rectify the issue can be damaging.
D. The database administrator is not the owner of the corporate address book application and therefore does not have authority to accept business risk.

R3-29 Which of the following is the **MOST** appropriate metric to measure how well the information security function is managing the administration of user access?

A. Elapsed time to suspend accounts of terminated users
B. Elapsed time to suspend accounts of users transferring
C. Ratio of actual accounts to actual end users
D. Percent of accounts with configurations in compliance

D is the correct answer.

Justification:
A. Elapsed time to suspend accounts of terminated users is only part of the picture and does not address the volume of requests.
B. Elapsed time to suspend accounts of users transferring is only part of the picture and does not address the volume of requests.
C. The ratio of actual accounts to actual end users does not indicate much in terms of how well security is administered.
D. **The percent of accounts with configurations in compliance is the best measure of how well the administration is being managed because this shows the overall impact.**

R3-30 Which of the following is the **PRIMARY** reason for periodically monitoring key risk indicators (KRIs)?

A. The cost of risk response needs to be minimized.
B. Errors in results of KRIs need to be minimized.
C. The risk profile may have changed.
D. Risk assessment needs to be continually improved.

C is the correct answer.

Justification:
A. Minimizing the cost of risk response efforts can be one of the outcomes, but this is not the primary reason.
B. If there are errors in results of KRIs, they can be minimized even without having periodic monitoring in place.
C. **The current set of risk impacting the enterprise can change over time and periodic monitoring of KRIs proactively identifies changes in the risk profile so that new risk can be addressed and changes in levels in existing risk can be better controlled.**
D. Risk assessment process improvements are not the reason for monitoring KRIs on a periodic basis.

R3-31 Which of the following is the **BEST** indicator of high maturity of an enterprise's IT risk management process?

A. People have appropriate awareness of risk and are comfortable talking about it.
B. Top management is prepared to invest more money in IT security.
C. Risk assessment is encouraged in all areas of IT and business management.
D. Business and IT are aligned in risk assessment and risk ranking.

A is the correct answer.

A. **Some of the most important measures of a mature IT risk management process are those related to a risk-aware culture—an enterprise where people recognize the risk inherent to their activities, are able to discuss it and are willing to work together to resolve the risk.**
B. While investment in IT security may strengthen the overall risk management posture of the enterprise, it is not an appropriate measure of IT risk management process maturity.
C. While risk assessment is an important step in the risk management process, it is not a good indicator of a mature risk management process, even when deployed across all business units and functions.
D. Alignment between IT and business is the foundation of an effective IT risk management process; however, it is not a good indicator of a mature IT risk management process.

DOMAIN 3—RISK MONITORING

R3-32 As part of risk monitoring, the administrator of a two-factor authentication system identifies a trusted independent source indicating that the algorithm used for generating keys has been compromised. The vendor of the authentication system has not provided further information. Which of the following is the **BEST** initial course of action?

A. Wait for the vendor to formally confirm the breach and provide a solution.
B. Determine and implement suitable compensating controls.
C. Identify all systems requiring two-factor authentication and notify their business owners.
D. Disable the system and rely on the single-factor authentication until further information is received.

C is the correct answer.

Justification:
A. Waiting on the vendor to acknowledge the vulnerability may result in an unacceptable exposure and may be considered negligent.
B. Determining suitable compensating controls is not appropriate without instructions from the responsible business owner.
C. **Business owners should be notified, even when some of the information may not be available. Business owners are responsible for responding to new risk.**
D. Disabling the system is not appropriate because there is no indication that the compromise will have an impact on the first-factor authentication.

R3-33 Which of the following **BEST** assists in the proper design of an effective key risk indicator (KRI)?

A. Generating the frequency of reporting cycles to report on the risk
B. Preparing a business case that includes the measurement criteria for the risk
C. Conducting a risk assessment to provide an overview of the key risk
D. Documenting the operational flow of the business from beginning to end

D is the correct answer.

Justification:
A. Generating the frequency of reporting for the KRI means nothing if the KRI is not designed.
B. A proper business case describes what is going to be done, why it is worth doing, how it will be accomplished and what resources will be required. It will not document the data points, structures or any other needed data for designing a KRI.
C. A risk assessment is the determination of a value of risk related to some situation and a recognized threat. While it contributes somewhat to the design of the KRI, there still is a need for additional information.
D. **Prior to starting to design the KRI, a risk manager must understand the end-to-end operational flow of the respective business. This gives insight into the detailed processes, data flows, decision-making process, acceptable levels of risk for the business, etc., which in turn give the risk manager the ability to apply top and bottom levels for the KRI.**

R3-34 The **MOST** likely trigger for conducting a comprehensive risk assessment is changes to:

A. the asset inventory.
B. asset classification levels.
C. the business environment.
D. information security policies.

C is the correct answer.

Justification:
A. Additions and removals of assets from the asset inventory is an ongoing process and will not generally trigger a risk assessment.
B. Due to risk assessment one can understand the classification requirements, but this is not the main trigger for which the actual risk assessment is performed periodically.
C. **Changes in the business environment in terms of new threats, vulnerabilities or changes to information assets deployment will act as a main trigger for conducting comprehensive risk assessment on a periodic basis. Based on periodic risk assessment, policies are modified rather than the other way around where risk assessment is performed based on policy changes already made.**
D. Changes to information security policies may occur when a risk assessment indicates deficiencies at the security policies level; changes to security policies do not trigger risk assessments.

DOMAIN 4—INFORMATION SYSTEMS CONTROL DESIGN AND IMPLEMENTATION (17%)

R4-1 Which of the following is the **MOST** important factor when designing IS controls in a complex environment?

A. Development methodologies
B. Scalability of the solution
C. Technical platform interfaces
D. Stakeholder requirements

D is the correct answer.

Justification:
A. Development methodologies are taken into consideration when designing IS controls to support stakeholder requirements.
B. Scalability of the solution is taken into consideration when designing IS controls to support stakeholder requirements.
C. Technical platform interfaces are taken into consideration when designing IS controls to support stakeholder requirements.
D. **The most important factor when designing IS controls is that they advance the interests of the business by addressing stakeholder requirements.**

R4-2 Investments in risk management technologies should be based on:

A. audit recommendations.
B. vulnerability assessments.
C. business climate.
D. value analysis.

D is the correct answer.

Justification:
A. Basing decisions on audit recommendations is reactive in nature and may not comprehensively address the key business needs.
B. Vulnerability assessments are useful, but they do not determine whether the cost is justified.
C. Demonstrated value takes precedence over the current business climate because the climate is ever changing.
D. **Investments in risk management technologies should be based on a value analysis and sound business case.**

R4-3 A global enterprise that is subject to regulation by multiple governmental jurisdictions with differing requirements should:

A. bring all locations into conformity with the aggregate requirements of all governmental jurisdictions.
B. bring all locations into conformity with a generally accepted set of industry best practices.
C. establish a baseline standard incorporating those requirements that all jurisdictions have in common.
D. establish baseline standards for all locations and add supplemental standards as required.

D is the correct answer.

Justification:
A. Seeking a lowest common denominator of requirements may cause certain locations to fail regulatory compliance.
B. Just using industry best practices may cause certain locations to fail regulatory compliance.
C. Forcing all locations to be in compliance with the regulations places an undue burden on those locations.
D. **It is more efficient to establish a baseline standard and then develop additional standards for locations that must meet specific requirements.**

DOMAIN 4—INFORMATION SYSTEMS CONTROL DESIGN AND IMPLEMENTATION

R4-4 Which of the following is **MOST** useful in developing a series of recovery time objectives (RTOs)?

A. Regression analysis
B. Risk analysis
C. Gap analysis
D. Business impact analysis (BIA)

D is the correct answer.

Justification:
A. Regression analysis is used to test changes to program modules.
B. Risk analysis is the process by which frequency and impact of risk scenarios are estimated; it is a component of a BIA.
C. Gap analysis is useful in addressing the differences between the current state and an ideal future state.
D. **RTOs are a primary deliverable of a BIA. RTOs relate to the financial impact of a system not being available.**

R4-5 The person responsible for ensuring that information is classified is the:

A. security manager.
B. technology group.
C. data owner.
D. senior management.

C is the correct answer.

Justification:
A. The security manager is responsible for applying security protection relative to the level of classification specified by the owner.
B. The technology group is delegated custody of the data by the data owner, but the group does not classify the information.
C. **The data owner is responsible for applying the proper classification to the data.**
D. Senior management is ultimately responsible for the enterprise.

R4-6 When transmitting personal information across networks, there **MUST** be adequate controls over:

A. encrypting the personal information.
B. obtaining consent to transfer personal information.
C. ensuring the privacy of the personal information.
D. change management.

C is the correct answer.

Justification:
A. Encryption is a method of achieving the actual control, but controls over the devices may not ensure adequate privacy protection. Therefore, encryption is a partial answer.
B. Consent is one of the protections that are frequently, but not always, required.
C. **Privacy protection is necessary to ensure that the receiving party has the appropriate level of protection of personal data.**
D. Change management is a core control that ensures that the privacy protections, encryption settings and consent processes are implemented as management intended; however, it will not directly address the privacy of the individuals.

DOMAIN 4—INFORMATION SYSTEMS CONTROL DESIGN AND IMPLEMENTATION

R4-7 Which of the following **BEST** addresses the risk of data leakage?

A. Incident response procedures
B. File backup procedures
C. Acceptable use policies (AUPs)
D. Database integrity checks

C is the correct answer.

Justification:
A. Confidentiality of information is not addressed by this choice.
B. Confidentiality of information is not addressed by this choice.
C. AUPs are the best measure for preventing the unauthorized disclosure of confidential information.
D. Confidentiality of information is not addressed by this choice.

R4-8 Which of the following measures is **MOST** effective against insider threats to confidential information?

A. Audit trail monitoring
B. A privacy policy
C. Role-based access control (RBAC)
D. Defense in depth

C is the correct answer.

Justification:
A. Audit trail monitoring is a detective control, which is "after the fact."
B. A privacy policy is not relevant to this risk.
C. RBAC provides access according to business needs; therefore, it reduces unnecessary access rights and enforces accountability.
D. The primary focus of defense in depth is external threats.

R4-9 Which of the following is the **BEST** way to ensure that contract programmers comply with organizational security policies?

A. Have the contractors acknowledge the security policies in writing.
B. Perform periodic security reviews of the contractors.
C. Explicitly refer to contractors in the security standards.
D. Create penalties for noncompliance in the contracting agreement.

B is the correct answer.

Justification:
A. Written acknowledgements of security policies do not help detect the failure of contract programmers to comply.
B. Periodic reviews are the most effective way of obtaining compliance.
C. Referring to the contract programs within security standards does not help detect the failure of contract programmers to comply with organizational security policies. It may establish responsibility for a control implementation and maintenance, but the control ownership and accountability remains within the enterprise itself.
D. Penalties do not help detect failure of contract programmers to comply with organizational security policies and can only be enforced once they are detected either by an audit or an incident.

DOMAIN 4—INFORMATION SYSTEMS CONTROL DESIGN AND IMPLEMENTATION

R4-10 Which of the following devices should be placed within a demilitarized zone (DMZ)?

A. An authentication server
B. A mail relay
C. A firewall
D. A router

B is the correct answer.

Justification:
A. An authentication server, due to its sensitivity, should always be placed on the internal network, never on a DMZ that is subject to compromise.
B. **A mail relay should normally be placed within a DMZ to shield the internal network.**
C. Firewalls may bridge a DMZ to another network, but do not technically reside within the DMZ network segment.
D. Routers may bridge a DMZ to another network, but do not technically reside within the DMZ network segment.

R4-11 Which of the following controls within the user provision process **BEST** enhance the removal of system access for contractors and other temporary users when it is no longer required?

A. Log all account usage and send it to their manager.
B. Establish predetermined, automatic expiration dates.
C. Ensure that each individual has signed a security acknowledgement.
D. Require managers to email security when the user leaves.

B is the correct answer.

Justification:
A. Logging is a detective control and, thus, is not as effective as the protective control of preexpiring user accounts.
B. **Predetermined expiration dates are the most effective means of removing systems access for temporary users.**
C. Requiring each individual to sign a security acknowledgement has little effect in this case.
D. Managers cannot be relied on to promptly send in termination notices.

R4-12 Which of the following **BEST** provides message integrity, sender identity authentication and nonrepudiation?

A. Symmetric cryptography
B. Message hashing
C. Message authentication code
D. Public key infrastructure (PKI)

D is the correct answer.

Justification:
A. Symmetric cryptography provides confidentiality.
B. Hashing can provide integrity and confidentiality.
C. Message authentication codes provide integrity.
D. **PKI combines public key encryption with a trusted third party to publish and revoke digital certificates that contain the public key of the sender. Senders can digitally sign a message with their private key and attach their digital certificate (provided by the trusted third party). These characteristics allow senders to provide authentication, integrity validation and nonrepudiation.**

DOMAIN 4—INFORMATION SYSTEMS CONTROL DESIGN AND IMPLEMENTATION

R4-13 Which of the following will **BEST** prevent external security attacks?

A. Securing and analyzing system access logs
B. Network address translation
C. Background checks for temporary employees
D. Static Internet protocol (IP) addressing

B is the correct answer.

Justification:
A. Securing and analyzing system access logs is a detective control.
B. **Network address translation is helpful by having internal addresses that are nonroutable.**
C. Background checks of temporary employees are more likely to prevent an attack launched from within the enterprise.
D. Static IP addressing does little to prevent an attack.

R4-14 Which of the following is the **BEST** control for securing data on mobile universal serial bus (USB) drives?

A. Authentication
B. Prohibiting employees from copying data to USB devices
C. Encryption
D. Limiting the use of USB devices

C is the correct answer..

Justification:
A. Authentication is the act of verifying identity of a user, system, service, etc. Authentication is generally coupled with identification and the delivery of access privileges; by itself, it is not a strong security measure.
B. Prohibiting employees from copying data to USB devices is a directive control that is challenging to enforce. It is not a strong security measure.
C. **Encryption provides the most effective protection of data on mobile devices.**
D. Limiting the use of USB devices is a preventive control that is challenging to enforce. It is not a strong security measure.

DOMAIN 4—INFORMATION SYSTEMS CONTROL DESIGN AND IMPLEMENTATION

R4-15 When configuring a biometric access control system that protects a high-security data center, the system's sensitivity level should be set to:

A. a lower equal error rate (EER).
B. a higher false acceptance rate (FAR).
C. a higher false reject rate (FRR).
D. the crossover error rate exactly.

C is the correct answer.

Justification:
A. When adjusting the sensitivity of a biometric access control system, the values for FAR and FRR adjust inversely (as indicated in the graph). At one point, the two values intersect and are equal. This condition creates the equal error rate (also called crossover rate), which is a measure of system accuracy.

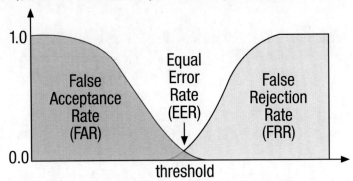

B. A higher false acceptance rate (FAR) is not desirable for a biometric system protecting a high-security data center because it is prone to falsely granting access to unauthorized individuals.
C. **Biometric access control systems are not infallible. When tuning the solution for a high-security data center, the sensitivity level should be adjusted to give preference either to an FRR (type I error rate) in which the system will be more prone to falsely reject access to a valid user than falsely granting access to an invalid user. In a very sensitive system, it may be desirable to minimize the number of false accepts—the number of unauthorized people allowed access. To do this, the system is tuned to be more sensitive, which causes the false rejects to increase—the number of authorized people not allowed access.**
D. While the crossover error rate is a measure of system accuracy, the best solution for a biometric access control system that protects a high-security data center is to err on the side of false rejects.

R4-16 Which of the following is the **MOST** effective measure to protect data held on mobile computing devices?

A. Protection of data being transmitted
B. Encryption of stored data
C. Power-on passwords
D. Biometric access control

B is the correct answer.

Justification:
A. While protecting data during transmission is important, it does not protect the data stored on the mobile device.
B. **Encryption of stored data will help ensure that the actual data cannot be recovered without the encryption key.**
C. Power-on passwords do not protect data effectively.
D. Biometric access control does not necessarily protect stored data.

DOMAIN 4—INFORMATION SYSTEMS CONTROL DESIGN AND IMPLEMENTATION

R4-17 Which of the following is **MOST** useful in managing increasingly complex deployments?

A. Policy development
B. A security architecture
C. Senior management support
D. A standards-based approach

B is the correct answer.

Justification:
A. Although policies guide direction, they do not effectively enable complex deployments.
B. **Deploying complex security initiatives and integrating a range of diverse projects and activities is more easily managed with the overview and relationships provided by a security architecture.**
C. Senior management support is important, yet is insufficient to ensure deployment.
D. Although standards may provide metrics for deployment, they do not effectively enable complex deployments.

R4-18 Business continuity plans (BCPs) should be written and maintained by:

A. the information security and information technology functions.
B. representatives from all functional units.
C. the risk management function.
D. executive management.

B is the correct answer.

Justification:
A. Although the information security and information technology functions areas have primary responsibility for disaster recovery planning, the BCP is a primary input representing the business priorities for IT/IS to build, test and maintain disaster recovery plans (DRPs).
B. **Business continuity planning is an enterprisewide activity; it is only successful if all business owners collaborate in the development, testing and maintenance of the plan.**
C. In many enterprises, risk management may oversee the business continuity program, but they are not in the best position to write or maintain business portions of the plan.
D. Executive management are responsible for assuring that the appropriate planning is performed, the plan is viable and understanding their responsibility should the plan be executed.

R4-19 Which of the following is a control designed to prevent segregation of duties (SoD) violations?

A. Enabling IT audit trails
B. Implementing two-way authentication
C. Reporting access log violations
D. Implementing role-based access

D is the correct answer.

Justification:
A. This choice is not correct because IT audits are detective controls and cannot ensure prevention of SoD violations.
B. This choice is not correct since two-way authentication ensures that client and server authenticate each other and does not help prevent SoD violations.
C. Reporting access log violations is a detective control and cannot ensure prevention of SoD violations.
D. **Implementing role-based access is a preventive method to mitigate SoD violations. All access levels can be adjusted according to the current role of the user, thus avoiding approvals of self-initiated transactions.**

DOMAIN 4—INFORMATION SYSTEMS CONTROL DESIGN AND IMPLEMENTATION

R4-20 System backup and restore procedures can **BEST** be classified as:

A. Technical controls
B. Detective controls
C. Corrective controls
D. Deterrent controls

C is the correct answer.

Justification:
A. Technical controls are safeguards incorporated into computer hardware, software or firmware. Operational procedures are nontechnical controls.
B. Detective controls help identify and escalate violations or attempted violations of security policy; examples include audit trails, intrusion detection tools and checksums.
C. **Corrective controls remediate vulnerabilities. If a system suffers harm so extensive that processing cannot continue, backup restore procedures enable that system to be recovered. This is a corrective measure that remediates the vulnerability of that system.**
D. Deterrent controls provide warnings that can discourage potential compromise; examples include warning banners or login screens.

R4-21 A business impact analysis (BIA) is **PRIMARILY** used to:

A. estimate the resources required to resume and return to normal operations after a disruption.
B. evaluate the impact of a disruption to an enterprise's ability to operate over time.
C. calculate the likelihood and impact of known threats on specific functions.
D. evaluate high-level business requirements.

B is the correct answer.

Justification:
A. Determining the resource requirements to resume and return to normal operations is part of business continuity planning.
B. **A BIA is primarily used for evaluating the impact of a disruption over time to an enterprise's ability to operate. It determines the urgency of each business activity. Key deliverables include recovery time objectives (RTOs) and recovery point objectives (RPOs).**
C. Likelihood and impact are calculated during risk analysis.
D. High-level business requirements are defined during the early phases of a system development life cycle (SDLC), not as part of a BIA.

DOMAIN 4—INFORMATION SYSTEMS CONTROL DESIGN AND IMPLEMENTATION

R4-22 Which of the following system development life cycle (SDLC) stages is **MOST** suitable for incorporating internal controls?

A. Development
B. Testing
C. Implementation
D. Design

D is the correct answer.

Justification:
A. Internal control requirements should be incorporated during development; however, unless the team already started incorporating internal controls during the preceding design phase, the project may incur a rework cost and is likely to affect project deliverables, project cost and the project time line.
B. Incorporating internal control requirements as late as the testing stage is likely to adversely affect project deliverables, project cost and the project time line.
C. Incorporating internal control requirements as late as the implementation stage is too late and may pose significant risk to the enterprise.
D. **Internal controls should be incorporated in the new system development at the earliest stage possible, i.e., at the design stage.**

R4-23 An enterprise has outsourced personnel data processing to a supplier, and a regulatory violation occurs during processing. Who will be held legally responsible?

A. The supplier, because it has the operational responsibility
B. The enterprise, because it owns the data
C. The enterprise and the supplier
D. The supplier, because it did not comply with the contract

B is the correct answer.

Justification:
A. The supplier has the operational responsibility pursuant to the contractual terms, but the regulatory authority will hold the customer responsible.
B. **The enterprise retains responsibility for the management of, and adherence to, policies, procedures and regulatory requirements. If the supplier fails to provide appropriate controls and/or performance based on the contract terms, the enterprise may have legal recourse. However, the regulatory authority will generally hold the enterprise responsible for failure to comply with regulations, including any penalties that may result.**
C. From the point of view of the regulatory authority the enterprise is legally responsible; in other words, the enterprise may be litigated and/or penalized for not fulfilling the contractual terms.
D. The supplier has the operational responsibility pursuant to the contractual terms, but the regulatory authority will hold the enterprise responsible.

DOMAIN 4—INFORMATION SYSTEMS CONTROL DESIGN AND IMPLEMENTATION

R4-24 Which of the following provides the formal authorization on user access?

A. Database administrator
B. Data owner
C. Process owner
D. Data custodian

B is the correct answer.

Justification:
A. The database administrator is responsible for overall database maintenance, support and performance and may grant access to data within the database once the data owner has approved the access request.
B. **The data owner provides the formal authorization to provide access to any user request.**
C. The process owner is responsible for a specific business process.
D. The data custodian is responsible for the safe custody, transport and storage of data and implementation of business rules, such as granting access to data, once the data owner has approved the access request.

R4-25 To determine the level of protection required for securing personally identifiable information, a risk practitioner should **PRIMARILY** consider the information:

A. source.
B. cost.
C. sensitivity.
D. validity.

C is the correct answer.

Justification:
A. The level of protection required is independent of the source of information and has more to do with the validity and information reliability than with protection.
B. The cost incurred to procure the information can partly determine the level of protection and in some way strengthens the information sensitivity factor, but cannot be the sole factor to determine protection level.
C. **Sensitivity of the information is the correct answer because the sensitive nature of the information takes precedence over source, cost or reliability.**
D. Validity of information does not dictate level of protection because protection is based on the content categorization.

R4-26 Which of the following provides the **GREATEST** level of information security awareness?

A. Job descriptions
B. A security manual
C. Security training
D. An organizational diagram

C is the correct answer.

Justification:
A. While job descriptions are useful to describe job-related roles and responsibilities, including those related to security, they do not provide sufficient detail to enable employees who are not already proficient in security to understand how they can actively support and contribute to a risk-aware culture. Such detail is generally only provided through security training.
B. A security manual is written for a technical audience and is usually not accessible to all staff. For this reason it is not a viable mechanism to make employees aware of their security responsibilities.
C. **Security training is the best way to inform all employees about information security awareness.**
D. An organizational diagram shows the various departmental hierarchies, but it is not associated to information security awareness. It can, however, be used by the information security team to determine which individuals need what type of information security awareness.

DOMAIN 4—INFORMATION SYSTEMS CONTROL DESIGN AND IMPLEMENTATION

R4-27 Risk assessments are **MOST** effective in a software development organization when they are performed:

A. before system development begins.
B. during system deployment.
C. during each stage of the system development life cycle (SDLC).
D. before developing a business case.

C is the correct answer.

Justification:
A. Performing a risk assessment before system development does not reveal any of the vulnerabilities introduced during development.
B. Performing a risk assessment at system deployment is not cost effective.
C. **Performing risk assessments at each stage of the SDLC is the most cost-effective way because it ensures that flaws are caught as soon as they occur.**
D. Performing a risk assessment before developing a business case does not reveal any of the vulnerabilities found during the SDLC.

R4-28 Management wants to ensure that IT is successful in delivering against business requirements. Which of the following **BEST** supports that effort?

A. An internal control system or framework
B. A cost-benefit analysis
C. A return on investment (ROI) analysis
D. A benchmark process

A is the correct answer.

Justification:
A. **For IT to be successful in delivering against business requirements, management should develop an internal control system that will make a link to the business requirements.**
B. A cost-benefit analysis, although useful, is not the most important element to align IT to business.
C. An ROI analysis is just one of the metrics to measure success of IT investments.
D. A benchmark process comes into place once a sound internal control framework has been enabled.

R4-29 Security technologies should be selected **PRIMARILY** on the basis of their:

A. evaluation in security publications.
B. compliance with industry standards.
C. ability to mitigate risk to organizational objectives.
D. cost compared to the enterprise's IT budget.

C is the correct answer.

Justification:
A. Evaluation in security publications is a valuable reference point when selecting a security technology, yet is secondary to the technology's ability to mitigate risk to the enterprise.
B. Compliance with industry standards may be an important aspect of selecting a security technology, but is secondary to the technology's ability to mitigate risk to the enterprise.
C. **The most fundamental evaluation criterion for the selection of security technology is its ability to reduce risk.**
D. While the cost of the technology in comparison to the budget is an important aspect for the selection of a suitable technology, it is secondary to the technology's ability to mitigate risk to the enterprise.

DOMAIN 4—INFORMATION SYSTEMS CONTROL DESIGN AND IMPLEMENTATION

R4-30 Which of the following groups would be the **MOST** effective in managing and executing an organization's risk program?

A. Midlevel management
B. Senior management
C. Frontline employees
D. The incident response team

A is the correct answer.

Justification:
A. **Midlevel management staff are the best to manage and execute an organization's risk management program because they are the most centrally located within the organizational hierarchy and they combine a sufficient breadth of influence with adequate proximity to day-to-day operations.**
B. Senior management staff are at too high a level to manage and execute the program, but their support is essential.
C. Frontline employees do not hold enough power and influence to manage and execute the program.
D. The incident response team may manage and execute a small portion of the program (incident response), but not the entire risk management program.

R4-31 Strong authentication is:

A. an authentication technique formally approved by a standardization organization.
B. the simultaneous use of several authentication techniques, e.g., password and badge.
C. an authentication system that makes use of cryptography.
D. an authentication system that uses biometric data to identify a person, i.e., a fingerprint.

B is the correct answer.

Justification:
A. The use of a standardized authentication technique in itself does not imply strong authentication.
B. **Authentication is the process of proving to someone that you are who you say you are—a guarantee of the sender's identity or origin. Because a third party vouches for the sender's identity, the recipient can rely on the authenticity of any transaction or message signed by that user. *Strong authentication* requires both something *you know* AND either something *you have* or are.**

   Three classic methods of authentication are:
   • **Something you *know*—passwords, the combination to a safe**
   • **Something you *have*—keys, tokens, badges**
   • **Something you *are*—physical traits, such as fingerprints, signature, iris pattern, keystroke patterns**
C. Cryptography is the practice and study of hiding information and—in relation to authentication—is mostly used to protect information that may be misused by a third party, such as passwords.
D. Biometrics consists of methods for uniquely recognizing humans based on one or more intrinsic physical or behavioral traits; it can help strengthen the authentication processes, but in itself does not imply strong authentication.

DOMAIN 4—INFORMATION SYSTEMS CONTROL DESIGN AND IMPLEMENTATION

R4-32 Which of the following is the **BIGGEST** concern for a chief information security officer (CISO) regarding interconnections with systems outside of the enterprise?

A. Requirements to comply with each other's contractual security requirements
B. Uncertainty that the other system will be available as needed
C. The ability to perform risk assessments on the other system
D. Ensuring that communication between the two systems is encrypted through a virtual private network (VPN) tunnel

A is the correct answer.

Justification:

A. **Ensuring that both systems comply with the contractual security requirements of both entities should be the primary concern of the risk practitioner. If one system falls out of compliance, then they both most likely will miss their respective security requirements.**
B. Uncertainty of the other system's availability is probably the primary concern of the business owner and users, not of the CISO.
C. The ability to perform risk assessment on the other system may or may not be a concern based on the interconnection agreement between the two systems.
D. Communication between the two systems may not necessarily require a VPN tunnel, or encryption. That requirement will be based on type of data being transmitted.

SCENARIO 2

A scenario is a mini-case study that describes a situation or an organization and requires candidates to answer one or more questions based on the information provided. A scenario can focus on a specific domain or on several domains. The CRISC exam will include scenarios.

QUESTIONS R4-33 THROUGH R4-34 REFER TO THE FOLLOWING INFORMATION:

The board of directors of a one-year-old start-up company has asked their chief information officer (CIO) to create all of the enterprise's IT policies and procedures, which will be managed and approved by the IT steering committee. The IT steering committee will make all of the IT decisions for the enterprise, including those related to the technology budget.

R4-33 The IT steering committee will be **BEST** represented by:

A. members of the executive board.
B. high-level members of the IT department.
C. IT experts from outside of the enterprise.
D. key members from each department.

D is the correct answer.

Justification:
A. If the steering committee is comprised of only the executive board, then it is likely that all of the goals will be high level and it is impractical for daily IT decisions to be made by the executive board.
B. If the steering committee is comprised of only the IT department, then it is likely that business objectives will be ignored in favor of technical best practices.
C. If the steering committee is comprised of only experts from outside of the enterprise, then business objectives will likely be ignored.
D. **The IT steering committee should be comprised of individuals from each department to ensure that the entire enterprise is represented and that all business objectives are more likely to be met.**

DOMAIN 4—INFORMATION SYSTEMS CONTROL DESIGN AND IMPLEMENTATION

SEE INFORMATION PRECEDING QUESTION R4-33

R4-34 Which type of IT organizational structure does the enterprise have?

A. Project-based
B. Centralized
C. Decentralized
D. Divisional

B is the correct answer.

Justification:
A. This choice is incorrect because a project-based enterprise is one where a group is formed temporarily to work on one particular project. The group that the CIO is setting up, and steering committees in general, is not temporary.
B. **With a centralized IT organizational structure, all of the decisions are made by one group for the entire enterprise.**
C. This choice is incorrect because with a decentralized organizational structure decisions are made by each division (sales, human resources [HR], etc). of the organization. If this were to occur, then there would be several different, and perhaps conflicting, IT policies.
D. This choice is incorrect because with a divisional organizational structure each geographic area or product or service will have its own group.

DOMAIN 5—INFORMATION SYSTEMS CONTROL MONITORING AND MAINTENANCE (18%)

R5-1 One way to determine control effectiveness is by determining:

A. the test results of intended objectives.
B. whether it is preventive, detective or compensatory.
C. the capability of providing notification of failure.
D. the evaluation and analysis of reliability.

A is the correct answer.

Justification:
A. **Control effectiveness requires a process to verify that the control process worked as intended. Examples such as dual-control or dual-entry bookkeeping provide verification and assurance that the process operated as intended.**
B. The type of control is not relevant.
C. Notification of failure is not determinative of control strength.
D. Reliability is not an indication of control strength; weak controls can be highly reliable, even if they are ineffective controls.

R5-2 Which of the following is the MOST effective way to ensure that third-party providers comply with the enterprise's information security policy?

A. Security awareness training
B. Penetration testing
C. Service level monitoring
D. Periodic auditing

D is the correct answer.

Justification:
A. Training can increase user awareness of the information security policy, but is not more effective than auditing.
B. Penetration testing can identify security vulnerability, but cannot ensure information compliance.
C. Service level monitoring can only pinpoint operational issues in the enterprise's operational environment.
D. **A regular audit exercise can spot any gap in the information security compliance.**

R5-3 Information security procedures should:

A. be updated frequently as new software is released.
B. underline the importance of security governance.
C. define the allowable limits of behavior.
D. describe security baselines for each platform.

A is the correct answer.

Justification:
A. **Often, security procedures have to change frequently to keep up with changes in software. Because a procedure is a how-to document, it must be kept up-to-date with frequent changes in software.**
B. High-level objectives of an enterprise, such as security governance, are normally addressed in a security policy.
C. Security policies define behavioral limits and are generally not updated as frequently as procedures.
D. Security standards define platform baselines; however, they do not provide the detail on how to apply the security baseline and are generally not updated as frequently as procedures.

DOMAIN 5—INFORMATION SYSTEMS CONTROL MONITORING AND MAINTENANCE

R5-4 Which of the following is used to determine whether unauthorized modifications were made to production programs?

A. An analytical review
B. Compliance testing
C. A system log analysis
D. A forensic analysis

B is the correct answer.

Justification:
A. Analytical review assesses the general control environment of an enterprise.
B. **Compliance testing helps to verify that the change management process has been applied consistently.**
C. It is unlikely that the system log analysis would provide information about the modification of programs.
D. Forensic analysis is a specialized technique for criminal investigation.

R5-5 Implementing continuous monitoring controls is the **BEST** option when:

A. legislation requires strong information security controls.
B. incidents may have a high impact and frequency.
C. incidents may have a high impact, but low frequency.
D. electronic commerce (e-commerce) is a primary business driver.

B is the correct answer.

Justification:
A. Regulations and legislation that require tight IT security measures focus on requiring enterprises to establish an IT security governance structure that manages IT security with a risk-based approach, so each organization decides which kinds of controls are implemented. Continuous monitoring is not necessarily a requirement.
B. **Because they are expensive, continuous monitoring control initiatives are used in areas where the risk is at its greatest level. These areas have a high impact and frequency of occurrence.**
C. Measures such as contingency planning are commonly used when incidents rarely happen, but have a high impact each time they happen. Continuous monitoring is unlikely to be necessary.
D. Continuous control monitoring initiatives are not needed in all e-commerce environments. There are some e-commerce environments where the impact of incidents is not high enough to support the implementation of this kind of initiative.

R5-6 Which of the following metrics is the **MOST** useful in measuring the monitoring of violation logs?

A. Penetration attempts investigated
B. Violation log reports produced
C. Violation log entries
D. Frequency of corrective actions taken

A is the correct answer.

Justification:
A. **The most useful metric is one that measures the degree to which complete follow-through has taken place.**
B. Violation log reports are not indicative of whether investigative action was taken. The most useful metric is one that measures the degree to which complete follow-through has taken place.
C. Violation log entries are not indicative of whether investigative action was taken. The most useful metric is one that measures the degree to which complete follow-through has taken place.
D. Frequency of corrective actions taken is not indicative of whether investigative action was taken. The most useful metric is one that measures the degree to which complete follow-through has taken place.

DOMAIN 5—INFORMATION SYSTEMS CONTROL MONITORING AND MAINTENANCE

R5-7 The **BEST** time to perform a penetration test is after:

A. a high turnover in systems staff.
B. an attempted penetration has occurred.
C. various infrastructure changes are made.
D. an audit has reported control weaknesses.

C is the correct answer.

Justification:
A. Turnover in systems staff does not warrant a penetration test, although it may warrant a review of password change practices and configuration management.
B. Conducting a test after an attempted penetration is not as productive because an enterprise should not wait until it is attacked to test its defenses.
C. **Changes in the systems infrastructure are most likely to inadvertently introduce new exposures.**
D. Any exposure identified by an audit should be corrected before it would be appropriate to test.

R5-8 Which of the following is the **BEST** way to ensure that a corporate network is adequately secured against external attack?

A. Utilize an intrusion detection system (IDS).
B. Establish minimum security baselines.
C. Implement vendor recommended settings.
D. Perform periodic penetration testing.

D is the correct answer.

Justification:
A. An IDS may detect an attempted attack, but it will not confirm whether the perimeter is secure.
B. Minimum security baselines are beneficial, but they will not provide the level of assurance that is provided by penetration testing.
C. Applying vendor recommended settings is beneficial, but it will not provide the level of assurance that is provided by penetration testing.
D. **Penetration testing is the best way to ensure that perimeter security is adequate.**

R5-9 Which of the following is **MOST** important for measuring the effectiveness of a security awareness program?

A. Increased interest in focus groups on security issues
B. A reduced number of security violation reports
C. A quantitative evaluation to ensure user comprehension
D. An increased number of security violation reports

D is the correct answer.

Justification:
A. Focus groups may or may not provide meaningful feedback, but in and of themselves do not provide metrics.
B. A reduction in the number of violation reports may not be indicative of a high level of security awareness.
C. To judge the effectiveness of user awareness training, measurable testing is necessary to confirm user comprehension. However, comprehension of what needs to be done does not ensure that action is taken when necessary. The most effective indicator for measuring success of an awareness program is an increase in the number of violation reports by staff.
D. **Of the choices offered, an increase in the number of violation reports is the best indicator of a high level of security awareness. As with automated alerts, each security violation report needs to be assessed for validity.**

DOMAIN 5—INFORMATION SYSTEMS CONTROL MONITORING AND MAINTENANCE

R5-10 Which of the following should be in place before a black box penetration test begins?

A. A clearly stated definition of scope
B. Previous test results
C. Proper communication and awareness training
D. An incident response plan

A is the correct answer.

Justification:
A. **A clearly stated definition of scope ensures a proper understanding of risk and success criteria.**
B. Previous test results help define the scope.
C. Communication and awareness training are not a necessary requirement.
D. An incident response plan is not a necessary requirement. In fact, a penetration test could help promote the creation and execution of the incident response plan.

R5-11 Which of the following **BEST** assists a risk practitioner in measuring the existing level of development of risk management processes against their desired state?

A. A capability maturity model (CMM)
B. Risk management audit reports
C. A balanced scorecard (BSC)
D. Enterprise security architecture

A is the correct answer.

Justification:
A. **The CMM grades processes on a scale of 0 to 5, based on their maturity, and is commonly used by entities to measure their existing state and then to determine the desired one.**
B. Risk management audit reports offer a limited view of the current state of risk management.
C. A BSC enables management to measure the implementation of strategy and assists in its translation into action.
D. Enterprise security architecture explains the security architecture of an entity in terms of business strategy, objectives, relationships, risk, constraints and enablers and provides a business-driven and business-focused view of security architecture.

R5-12 An enterprise is hiring a consultant to help determine the maturity level of the risk management program. The **MOST** important element of the request for proposal (RFP) is the:

A. sample deliverable.
B. past experience of the engagement team.
C. methodology used in the assessment.
D. references from other organizations.

C is the correct answer.

Justification:
A. Sample deliverables only tell how the assessment is presented, not the process.
B. Past experience of the engagement team is not as important as the methodology used.
C. **Methodology illustrates the process and formulates the basis to align expectations and the execution of the assessment. This also provides a picture of what is required of all parties involved in the assessment.**
D. References from other organizations are important, but not as important as the methodology used in the assessment.

R5-13 A third party is engaged to develop a business application. Which of the following **BEST** measures for the existence of back doors?

A. Security code reviews for the entire application
B. System monitoring for traffic on network ports
C. Reverse engineering the application binaries
D. Running the application from a high-privileged account on a test system

A is the correct answer.

Justification:
A. **Security code reviews for the entire application are the best measure and involve reviewing the entire source code to detect all instances of back doors.**
B. System monitoring for traffic on network ports is not able to detect all instances of back doors, is time consuming and takes a lot of effort.
C. Reverse engineering the application binaries may not provide any definite clues.
D. Back doors do not surface by running the application on high-privileged accounts because back doors are usually hidden accounts in the applications.

R5-14 Which of the following techniques **BEST** helps determine whether there have been unauthorized program changes since the last authorized program update?

A. A test data run
B. An automated code comparison
C. A code review
D. A review of code migration procedures

B is the correct answer.

Justification:
A. Test data runs help verify the processing of preselected transactions, but provide no evidence about unexercised portions of a program.
B. **An automated code comparison is the process of comparing two versions of the same program to determine whether the two correspond. It is an efficient technique because it is an automated procedure.**
C. A code review is the process of reading program source code listings to determine whether the code contains potential errors or inefficient statements. A code review can be used as a means of code comparison, but it is not efficient.
D. A review of code migration procedures would not detect program changes.

DOMAIN 5—INFORMATION SYSTEMS CONTROL MONITORING AND MAINTENANCE

R5-15 A substantive test to verify that tape library inventory records are accurate is:

A. determining whether bar code readers are installed.
B. conducting a physical count of the tape inventory.
C. checking whether receipts and issues of tapes are accurately recorded.
D. determining whether the movement of tapes is authorized.

B is the correct answer.

Justification:
A. Testing the existence of bar code readers is a compliance test, not a substantive test. A substantive test includes gathering evidence to evaluate the integrity of individual transactions, data or other information.
B. **A substantive test includes gathering evidence to evaluate the integrity of individual transactions, data or other information. Conducting a physical count of the tape inventory is a substantive test.**
C. Confirming that receipts and issues of tapes are accurately recorded is a compliance test, not a substantive test. A substantive test includes gathering evidence to evaluate the integrity of individual transactions, data or other information.
D. Testing the approval of tape movements is a compliance test, not a substantive test. A substantive test includes gathering evidence to evaluate the integrity of individual transactions, data or other information.

R5-16 When a significant vulnerability is discovered in the security of a critical web server, immediate notification should be made to the:

A. development team to remediate.
B. data owners to mitigate damage.
C. system owner to take corrective action.
D. incident response team to investigate.

C is the correct answer.

Justification:
A. The development team may be called on by the system owner to resolve the vulnerability.
B. Data owners are notified only if the vulnerability could have compromised data.
C. **To correct the vulnerabilities, the system owner needs to be notified quickly, before an incident can take place.**
D. Investigation by the incident response team is not correct because the incident has not taken place and notification could delay implementation of the fix.

R5-17 The **BEST** method for detecting and monitoring a hacker's activities without exposing information assets to unnecessary risk is to utilize:

A. firewalls.
B. bastion hosts.
C. honeypots.
D. screened subnets.

C is the correct answer.

Justification:
A. Firewalls attempt to keep the hacker out, which is a preventive control.
B. Bastion hosts attempt to keep the hacker out, which is a preventive control.
C. **The best choice for diverting a hacker away from critical files and alerting security of the hacker's presence are honeypots, often referred to as decoy files.**
D. Screened subnets or demilitarized zones (DMZs) provide a middle ground between the trusted internal network and the external, untrusted Internet.

DOMAIN 5—INFORMATION SYSTEMS CONTROL MONITORING AND MAINTENANCE

R5-18 Which of the following is the **BEST** way to verify that critical production servers are utilizing up-to-date antivirus signature files?

A. Check a sample of servers.
B. Verify the date that signature files were last pushed out.
C. Use a recently identified benign virus to test whether it is quarantined.
D. Research the most recent signature file, and compare it to the console.

A is the correct answer.

Justification:
A. **The only effective way to check the currency of signature files is to look at a sample of servers.**
B. The fact that an update was pushed out to a server does not guarantee that it was properly loaded onto that server. In conjunction with the sample testing, the process for updating the signature files should be verified.
C. Personnel should never release a virus, no matter how benign.
D. Comparing the vendor's most recent signature file to the management console is not indicative of whether the file was properly loaded on the server.

R5-19 Security administration efforts are **BEST** reduced through the deployment of:

A. access control lists (ACLs).
B. discretionary access controls (DACs).
C. mandatory access controls (MACs).
D. role-based access controls (RBACs).

D is the correct answer.

Justification:
A. ACLs fall in the category of discretionary access controls in that they assign permissions to specific operations with meaning in the enterprise, and not to the low-level data objects. ACLs can never be as fine-tuned as RBACs and need frequent modification.
B. The traditional discretionary access control (DAC) model is based on resource ownership and access to resources is based on user identity. DAC requires changes to a user access profile every time that the identity changes.
C. In a mandatory access control (MAC) model, access to resources is based on an object's security level, while users are granted security clearance. Only administrators can modify an object's security label or a user's security clearance, and this increases the administrative overhead cost.
D. **RBACs tie individuals to specific roles. The use of roles, hierarchies and constraints to organize privileges reduces the security administration effort when individuals change positions. RBACs are also known as nondiscretionary access controls.**

DOMAIN 5—INFORMATION SYSTEMS CONTROL MONITORING AND MAINTENANCE

R5-20 Which of the following is the **BEST** metric to manage the information security program?

A. The number of systems that are subject to intrusion detection
B. The amount of downtime caused by security incidents
C. The time lag between detection, reporting and acting on security incidents
D. The number of recorded exceptions from the minimum information security requirements

D is the correct answer.

Justification:
A. The number of systems subject to intrusion has no relevance to the quality of security management, but has more to do with the enterprise's vulnerability.
B. The amount of downtime is a measure of the scale of the threat.
C. The time lag is a measure of the responsiveness of the security team.
D. **The number of exceptions from set requirements is a direct correlation to the quality of the security program.**

R5-21 Which of the following is the **BEST** approach when malicious code from a spear phishing attack resides on the network and the finance department is concerned that scanning the network will slow down work and delay quarter-end reporting?

A. Instruct finance to finalize quarter-end reporting, then perform a scan of the entire network.
B. Block all outgoing traffic to avoid outbound communication to the expecting command host.
C. Scan network devices that are not supporting financial reporting, and then scan the critical finance drives at night.
D. Perform a staff survey and ask staff to report if they are aware of the enterprise being a target of a spear phishing attack.

C is the correct answer.

Justification:
A. Instructing finance to finalize quarter-end reporting and not performing scanning until it is complete is not the most efficient approach to deal with a potential incident.
B. Blocking all outgoing traffic to avoid outbound communication to the expecting command host will also block justified outbound business communication and is not the most effective approach to deal with a potential incident.
C. **Implementing an incremental scanning approach helps confirm the potential risk while allowing the business unit responsible for financial reporting to conduct their operations with minimal interference.**
D. Asking staff if they are aware of being a target of a spear phishing attack is useless because spear phishing attacks are specifically designed to look like authentic email messages and users are not likely to be aware that they have been victimized.

R5-22 Which of the following **BEST** describes the objective of a business impact analysis (BIA)?

A. The identification of threats, risk and vulnerabilities that can adversely affect the enterprise
B. The development of procedures for initial response and stabilization of situations during an emergency
C. The identification of time-sensitive critical business functions and interdependencies
D. The development of communication procedures in the case of a crisis impacting the business

C is the correct answer.

Justification:
A. The identification of threats, risk and vulnerabilities is the objective of risk identification and analysis.
B. The development of procedures for initial response and stabilization of situations during an emergency is a key output of preparedness and response planning.
C. **Identification of time-sensitive critical business functions and interdependencies is a deliverable of the BIA; this is reflected partially by metrics, such as recovery time objectives (RTOs) and recovery point objectives (RPOs).**
D. Communication procedures are beneficial to every business process, including crisis management; however, they are not the main deliverable of the BIA and relate more closely to business continuity and disaster recovery planning.

DOMAIN 5—INFORMATION SYSTEMS CONTROL MONITORING AND MAINTENANCE

R5-23 Which of the following **MOST** effectively ensures that service provider controls are within the guidelines set forth in the organization's information security policy?

A. Service level monitoring
B. Penetration testing
C. Security awareness training
D. Periodic auditing

D is the correct answer.

Justification:
A. Service level monitoring helps pinpoint the service provider's operational issues, but is not designed to ensure compliance.
B. Penetration testing helps identify system vulnerabilities, but is not designed to ensure compliance.
C. Security awareness training is a preventive measure to increase user awareness of the information security policy, but is not designed to ensure compliance.
D. **Periodic audits help ensure compliance with the organization's information security policy.**

R5-24 Which of the following is the **BEST** option to ensure that corrective actions are taken after a risk assessment is performed?

A. Conduct a follow-up review.
B. Interview staff member(s) responsible for implementing the corrective action.
C. Ensure that an organizational executive documents that the corrective action was taken.
D. Run a monthly report and verify that the corrective action was taken.

A is the correct answer.

Justification:
A. **Conducting a follow-up review is correct because it is the only option that ensures that the corrective action was taken.**
B. Interviewing the staff member(s) is incorrect because there is no concrete proof that the action was taken.
C. Documenting that a corrective action was taken is incorrect because it does not ensure that the action was taken, but it is a good step to take once a follow-up audit is conducted.
D. A monthly report may not be specific enough to ensure that the corrective action was taken.

R5-25 Which of the following **BEST** ensures that appropriate mitigation occurs on identified information systems vulnerabilities?

A. Presenting root cause analysis to the management of the organization
B. Implementing software to input the action points
C. Incorporating the findings into the annual report to shareholders
D. Assigning action plans with deadlines to responsible personnel

D is the correct answer.

Justification:
A. Presenting findings to management will increase management awareness; however, it does not ensure that action will be taken by the staff.
B. Software can help in monitoring the progress of mitigations, but it will not ensure completeness.
C. Reporting to shareholders does not ensure that the mitigation will be completed.
D. **Assigning mitigation to personnel establishes responsibility for its completion within the deadline.**

DOMAIN 5—INFORMATION SYSTEMS CONTROL MONITORING AND MAINTENANCE

R5-26 The **MOST** important objective of regularly testing information system controls is to:

A. identify design flaws, failures and redundancies.
B. provide the necessary evidence to support management assertions.
C. assess the control risk and formulate an opinion on the level of reliability.
D. evaluate the need for a risk assessment and indicate the corrective action(s) to be taken, where applicable.

A is the correct answer.

Justification:
A. **This choice is the best statement because it contains the necessary activities needed to ensure that the control is designed correctly and is operating effectively and efficiently during the production phase.**
B. This activity is performed after the completion of an assessment or audit of the information system control.
C. This activity is primarily performed during the design phase of the information system control.
D. Risk assessments do not depend on testing of controls.

R5-27 The **MOST** effective starting point to determine whether an IT system continues to meet the enterprise's business objectives is to conduct interviews with:

A. executive management.
B. IT management.
C. business process owners.
D. external auditors.

C is the correct answer.

Justification:
A. Executive management will be able to provide the overall picture of the enterprise's business objectives.
B. IT management is important, but should not be the starting point because they likely do not see a clear picture of all organizational objectives or how the business plans to use IT in the future.
C. **Business process owners are an effective starting point for conducting interviews to ensure that IT systems are meeting their individual business process needs.**
D. External auditors can be useful for an objective view on control performance of the IT systems, but they are not a starting point in determining if an IT system continues to meet organizational objectives.

DOMAIN 5—INFORMATION SYSTEMS CONTROL MONITORING AND MAINTENANCE

R5-28 Despite a comprehensive security awareness program annually undertaken and assessed for all staff and contractors, an enterprise has experienced a breach through a spear phishing attack. What is the **MOST** effective way to improve security awareness?

A. Review the security awareness program and improve coverage of social engineering threats.
B. Launch a disciplinary process against the people who leaked the information.
C. Perform a periodic social engineering test against all staff and communicate summary results to the staff.
D. Implement a data loss prevention system that automatically points users to corporate policies.

C is the correct answer.

Justification:
A. The awareness program should be periodically reviewed, despite the incident. However, because the awareness program is comprehensive and undertaken by all staff, the review will not create the necessary improvement.
B. Spear phishing attacks are designed to look like justified communication from a trusted source; thus, a disciplinary process is inappropriate.
C. **Users who are aware of security threats may need a reminder that these threats are real. Periodic social engineering tests help in maintaining a level of alertness.**
D. Investment in the data loss prevention system may not be justified because it is designed to protect against a loss of special types of data and would not stop disclosure of user credentials.

R5-29 Which of the following **BEST** helps identify information systems control deficiencies?

A. Gap analysis
B. The current IT risk profile
C. The IT controls framework
D. Countermeasure analysis

A is the correct answer.

Justification:
A. **Controls are deployed to achieve the desired control objectives based on risk assessments and business requirements. The gap between desired control objectives and actual IS control design and operational effectiveness identifies IS control deficiencies.**
B. Without knowing the gap between desired state and current state, one cannot identify the control deficiencies.
C. The IT controls framework is a generic document with no information such as desired state of IS controls and current state of the enterprise; therefore, it will not help in identifying IS control deficiencies.
D. Countermeasure analysis only helps in identifying deficiencies in countermeasures and not in the full set of primary controls.

DOMAIN 5—INFORMATION SYSTEMS CONTROL MONITORING AND MAINTENANCE

R5-30 Which of the following **BEST** ensures that information systems control deficiencies are appropriately remediated?

A. A risk mitigation plan
B. Risk reassessment
C. Control risk reevaluation
D. Countermeasure analysis

A is the correct answer.

Justification:
A. **Once risk is identified due to current IS control deficiencies, a risk mitigation plan will have the set of controls with a detailed plan, including countermeasures that can best help in risk remediation to an appropriate level.**
B. Risk reassessment is required when there are major changes to the risk environment; it is usually performed after a period of time as defined by management.
C. Control risk reevaluation helps in further validation of IS control deficiencies, but does not ensure that these deficiencies are actually remediated.
D. Countermeasure analysis is targeted toward countermeasures and may provide some information on deficiencies, but does not ensure that IS control deficiencies are remediated.

R5-31 Which of the following is **MOST** critical when system configuration files for a critical enterprise application system are being reviewed?

A. Configuration files are frequently changed.
B. Changes to configuration files are recorded.
C. Access to configuration files is not restricted.
D. Configuration values do not impact system efficiency.

C is the correct answer.

Justification:
A. Changes to configuration files are based on business need and may be frequent. The key is validating that these changes are documented in the change management system and approved before being implemented into the production environment.
B. Even if the changes to the parameter file are recorded, this is less critical than access to configuration files because if access is not restricted, then the unauthorized user can disable recording of changes in the system using accounts with a high privilege.
C. **If access to configuration files is not restricted, then the security of the overall system will be in question.**
D. If access to the parameter file is not restricted, then the security of the overall system will be in question because access is bypassed; the system can be impacted in many ways and the efficiency of the system will be a lesser problem than losing control of the entire system.

DOMAIN 5—INFORMATION SYSTEMS CONTROL MONITORING AND MAINTENANCE

R5-32 When assessing the performance of a critical application server, the **MOST** reliable assessment results may be obtained from:

A. activation of native database auditing.
B. documentation of performance objectives.
C. continuous monitoring.
D. documentation of security modules.

C is the correct answer.

Justification:
A. Native database audit logs are a good detective control, but do not provide information about the application server performance.
B. Documentation of performance objectives is important, but does not provide information about the application server performance.
C. **It is essential to obtain monitoring data in a consistent manner to achieve reliable results. Changing the monitoring methodology frequently does not enable time-series data comparison.**
D. Documentation of associated security modules may be helpful, but does not provide information about the application server performance.

R5-33 The **PRIMARY** reason for developing an enterprise security architecture is to:

A. align security strategies between the functional areas of an enterprise and external entities.
B. build a barrier between the IT systems of an enterprise and the outside world.
C. help with understanding of the enterprise's technologies and the interactions between them.
D. protect the enterprise from external threats and proactively monitor the corporate network.

A is the correct answer.

Justification:
A. **The enterprise security architecture must align with the strategies and objectives of the enterprise, taking into consideration the importance of the free flow of information within an enterprise as well as business with partners, customers and suppliers.**
B. Building a barrier between the IT systems of an enterprise and the outside world without proper alignment with business, information and technology may interfere with valid business processes.
C. The development of enterprise security architecture should not only take into consideration every piece of technology that exists in the enterprise, but also provide an understanding of how and why these technologies interact with each other as well as outside processes, suppliers, partners, customers and existing business processes to achieve enterprise objectives.
D. An enterprise security architecture does not protect the enterprise from threats nor does it perform monitoring of threats; it lays down a blueprint, including internal and external controls needed to protect the enterprise.

DOMAIN 5—INFORMATION SYSTEMS CONTROL MONITORING AND MAINTENANCE

R5-34 The **PRIMARY** goal of a postincident review is to:

A. gather evidence for subsequent legal action.
B. identify ways to improve the response process.
C. identify individuals who failed to take appropriate action.
D. make a determination as to the identity of the attacker.

B is the correct answer.

Justification:
A. Evidence should already have been gathered earlier in the process.
B. The goal of a postincident review is to identify ways to improve the incident response process.
C. A postincident review should not focus on finding and punishing individuals who did not take appropriate action, but rather on establishing a process to reduce the likelihood of similar incidents in the future and to improve the incident response process.
D. Identification of the attacker is not an objective of the postincident review process.

R5-35 During an organizational risk assessment it is noted that many corporate IT standards have not been updated. The **BEST** course of action is to:

A. review the standards against current requirements and make a determination of adequacy.
B. determine that the standards should be updated annually.
C. report that IT standards are adequate and do not need to be updated.
D. review the IT policy document and see how frequently IT standards should be updated.

A is the correct answer.

Justification:
A. The risk practitioner should verify that the standards are still adequate. If standards are lacking, then they should be updated.
B. Standards may or may not need to be updated, but should be reviewed annually for adequacy.
C. The risk practitioner cannot report that the IT standards are accurate until they are reviewed.
D. Reviewing the IT policy will not help determine whether the standards are still adequate or relevant.

R5-36 There is an increase in help desk call levels because the vendor hosting the human resources (HR) self-service portal has reduced the password expiration from 90 to 30 days. The corporate password policy requires password expiration after 60 days and HR is unaware of the change. The risk practitioner should **FIRST**:

A. formally investigate the cause of the unauthorized change.
B. request the service provider reverse the password expiration period to 90 days.
C. initiate a request to strengthen the corporate password expiration requirement to 30 days.
D. notify employees of the change in password expiration period.

A is the correct answer.

Justification:
A. **The key risk for the business process owner is that the external vendor is performing unauthorized changes to the configuration settings. All other actions are incorrect, because any change carries risk and requires a rigorous management approach.**
B. Reversing the change to a 90-day password expiration period would result in noncompliance with corporate policy.
C. While changing the corporate password policy to be more stringent may seem more secure, corporate policy should always be driven by business requirements, not the actions of service providers. Effective security requirements balance security with operational functionality.
D. While exceeding the requirements of the corporate password policy may seem more secure, such activities should follow a formal change management process and not be driven by individual actions or arbitrary changes from a service provider. Effective security requirements balance security with operational functionality.

DOMAIN 5—INFORMATION SYSTEMS CONTROL MONITORING AND MAINTENANCE

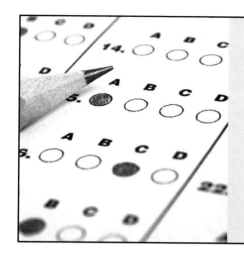

POSTTEST

If you wish to take a posttest to determine strengths and weaknesses, the Sample Exam begins on page 93 and the posttest answer sheet begins on page 127. You can score your posttest with the Sample Exam Answer and Reference Key on page 123.

SAMPLE EXAM

1. Which of the following controls within the user provision process **BEST** enhance the removal of system access for contractors and other temporary users when it is no longer required?

    A. Log all account usage and send it to their manager.
    B. Establish predetermined, automatic expiration dates.
    C. Ensure that each individual has signed a security acknowledgement.
    D. Require managers to email security when the user leaves.

2. Which of the following factors should be included when assessing the impact of losing network connectivity for 18 to 24 hours?

    A. The hourly billing rate charged by the carrier
    B. Financial losses incurred by affected business units
    C. The value of the data transmitted over the network
    D. An aggregate compensation of all affected business users

3. How often should risk be evaluated?

    A. Annually or when there is a significant change
    B. Once a year for each business process and subprocess
    C. Every three to six months for critical business processes
    D. Only after significant changes occur

4. There is an increase in help desk call levels because the vendor hosting the human resources (HR) self-service portal has reduced the password expiration from 90 to 30 days. The corporate password policy requires password expiration after 60 days and HR is unaware of the change. The risk practitioner should **FIRST**:

    A. formally investigate the cause of the unauthorized change.
    B. request the service provider reverse the password expiration period to 90 days.
    C. initiate a request to strengthen the corporate password expiration requirement to 30 days.
    D. notify employees of the change in password expiration period.

5. When assessing the performance of a critical application server, the **MOST** reliable assessment results may be obtained from:

    A. activation of native database auditing.
    B. documentation of performance objectives.
    C. continuous monitoring.
    D. documentation of security modules.

6. Which of the following risk assessment outputs is **MOST** suitable to help justify an organizational information security program?

    A. An inventory of risk that may impact the enterprise
    B. Documented threats to the enterprise
    C. Evaluation of the consequences
    D. A list of appropriate controls for addressing risk

SAMPLE EXAM

7. Which of the following is MOST beneficial to the improvement of an enterprise's risk management process?

   A. Key risk indicators (KRIs)
   B. External benchmarking
   C. The latest risk assessment
   D. A maturity model

8. Which of the following is MOST useful in developing a series of recovery time objectives (RTOs)?

   A. Regression analysis
   B. Risk analysis
   C. Gap analysis
   D. Business impact analysis (BIA)

9. Which of the following is the MOST important reason for conducting periodic risk assessments?

   A. Risk assessments are not always precise.
   B. Reviewers can optimize and reduce the cost of controls.
   C. Periodic risk assessments demonstrate the value of the risk management function to senior management.
   D. Business risk is subject to frequent change.

10. Which of the following BEST assists a risk practitioner in measuring the existing level of development of risk management processes against their desired state?

    A. A capability maturity model (CMM)
    B. Risk management audit reports
    C. A balanced scorecard (BSC)
    D. Enterprise security architecture

11. Which of the following measures is MOST effective against insider threats to confidential information?

    A. Audit trail monitoring
    B. A privacy policy
    C. Role-based access control (RBAC)
    D. Defense in depth

12. Risk assessments are MOST effective in a software development organization when they are performed:

    A. before system development begins.
    B. during system deployment.
    C. during each stage of the system development life cycle (SDLC).
    D. before developing a business case.

13. Which of the following BEST describes the role of management in implementing a risk management strategy?

    A. Ensure that the planning, budgeting and performance of information security components are appropriate.
    B. Assess and incorporate the results of the risk management activity into the decision-making process.
    C. Identify, evaluate and minimize risk to IT systems that support the mission of the organization.
    D. Understand the risk management process so that appropriate training materials and programs can be developed.

14. Which of the following provides the formal authorization on user access?

    A. Database administrator
    B. Data owner
    C. Process owner
    D. Data custodian

SAMPLE EXAM

15. The **PRIMARY** advantage of creating and maintaining a risk register is to:

    A. ensure that an inventory of potential risk is maintained.
    B. ensure that all assets have a low residual risk.
    C. define the risk assessment methodology.
    D. study a variety of risk and redefine the threat landscape.

16. The **MOST** likely trigger for conducting a comprehensive risk assessment is changes to:

    A. the asset inventory.
    B. asset classification levels.
    C. the business environment.
    D. information security policies.

17. Which of the following system development life cycle (SDLC) stages is **MOST** suitable for incorporating internal controls?

    A. Development
    B. Testing
    C. Implementation
    D. Design

18. Which of the following is minimized when acceptable risk is achieved?

    A. Transferred risk
    B. Control risk
    C. Residual risk
    D. Inherent risk

19. After a risk assessment study, a bank with global operations decided to continue doing business in certain regions of the world where identity theft is widespread. To **MOST** effectively deal with the risk, the business should:

    A. implement monitoring techniques to detect and react to potential fraud.
    B. make the customer liable for losses if the customer fails to follow the bank's advice.
    C. increase its customer awareness efforts in those regions.
    D. outsource credit card processing to a third party.

20. Which of the following is **MOST** useful in managing increasingly complex deployments?

    A. Policy development
    B. A security architecture
    C. Senior management support
    D. A standards-based approach

21. Which of the following groups would be the **MOST** effective in managing and executing an organization's risk program?

    A. Midlevel management
    B. Senior management
    C. Frontline employees
    D. The incident response team

CRISC Review Questions, Answers & Explanations Manual 2013
ISACA. All Rights Reserved.

SAMPLE EXAM

22. Which of the following is the **BEST** control for securing data on mobile universal serial bus (USB) drives?

    A. Authentication
    B. Prohibiting employees from copying data to USB devices
    C. Encryption
    D. Limiting the use of USB devices

23. Which of the following is the **BIGGEST** concern for a chief information security officer (CISO) regarding interconnections with systems outside of the enterprise?

    A. Requirements to comply with each other's contractual security requirements
    B. Uncertainty that the other system will be available as needed
    C. The ability to perform risk assessments on the other system
    D. Ensuring that communication between the two systems is encrypted through a virtual private network (VPN) tunnel

24. The **PRIMARY** focus of managing IT-related business risk is to protect:

    A. information.
    B. hardware.
    C. applications.
    D. databases.

25. A risk practitioner receives a message late at night that critical IT equipment will be delivered several days late due to flooding. Fortunately, a reciprocal agreement exists with another company for a replacement until the equipment arrives. This is an example of risk:

    A. transfer.
    B. avoidance.
    C. acceptance.
    D. mitigation.

26. Which of the following is **MOST** effective in assessing business risk?

    A. A use case analysis
    B. A business case analysis
    C. Risk scenarios
    D. A risk plan

27. Which of the following provides the **BEST** view of risk management?

    A. An interdisciplinary team
    B. A third-party risk assessment service provider
    C. The enterprise's IT department
    D. The enterprise's internal compliance department

28. A procurement employee notices that new printer models offered by the vendor keep a copy of all printed documents on a built-in hard disk. Considering the risk of unintentionally disclosing confidential data, the employee should:

    A. proceed with the order and configure printers to automatically wipe all the data on disks after each print job.
    B. notify the security manager to conduct a risk assessment for the new equipment.
    C. seek another vendor that offers printers without built-in hard disk drives.
    D. procure printers with built-in hard disks and notify staff to wipe hard disks when decommissioning the printer.

SAMPLE EXAM

29. A database administrator notices that the externally hosted, web-based corporate address book application requires users to authenticate, but that the traffic between the application and users is not encrypted. The **MOST** appropriate course of action is to:

    A. notify the business owner and the security manager of the discovery and propose an addition to the risk register.
    B. contact the application administrators and request that they enable encryption of the application's web traffic.
    C. alert all staff about the vulnerability and advise them not to log on from public networks.
    D. accept that current controls are suitable for nonsensitive business data.

30. Which of the following leads to the **BEST** optimal return on security investment?

    A. Deploying maximum security protection across all of the information assets
    B. Focusing on the most important information assets and then determining their protection
    C. Deploying minimum protection across all the information assets
    D. Investing only after a major security incident is reported to justify investment

31. During an organizational risk assessment it is noted that many corporate IT standards have not been updated. The **BEST** course of action is to:

    A. review the standards against current requirements and make a determination of adequacy.
    B. determine that the standards should be updated annually.
    C. report that IT standards are adequate and do not need to be updated.
    D. review the IT policy document and see how frequently IT standards should be updated.

32. Which of the following provides the **GREATEST** support to a risk practitioner recommending encryption of corporate laptops and removable media as a risk mitigation measure?

    A. Benchmarking with peers
    B. Evaluating public reports on encryption algorithm in the public domain
    C. Developing a business case
    D. Scanning unencrypted systems for vulnerabilities

33. Risk assessments should be repeated at regular intervals because:

    A. omissions in earlier assessments can be addressed.
    B. periodic assessments allow various methodologies.
    C. business threats are constantly changing.
    D. they help raise risk awareness among staff.

34. Which of the following is the **GREATEST** risk of a policy that inadequately defines data and system ownership?

    A. Audit recommendations may not be implemented.
    B. Users may have unauthorized access to originate, modify or delete data.
    C. User management coordination does not exist.
    D. Specific user accountability cannot be established.

35. Which of the following documents **BEST** identifies an enterprise's compliance risk and the corrective actions in progress to meet these regulatory requirements?

    A. An internal audit report
    B. A risk register
    C. An external audit report
    D. A risk assessment report

SAMPLE EXAM

36. Which of the following BEST ensures that appropriate mitigation occurs on identified information systems vulnerabilities?

    A. Presenting root cause analysis to the management of the organization
    B. Implementing software to input the action points
    C. Incorporating the findings into the annual report to shareholders
    D. Assigning action plans with deadlines to responsible personnel

37. Which of the following provides the GREATEST level of information security awareness?

    A. Job descriptions
    B. A security manual
    C. Security training
    D. An organizational diagram

38. Whether a risk has been reduced to an acceptable level should be determined by:

    A. IS requirements.
    B. information security requirements.
    C. international standards.
    D. organizational requirements.

39. Information that is no longer required to support the main purpose of the business from an information security perspective should be:

    A. analyzed under the retention policy.
    B. protected under the information classification policy.
    C. analyzed under the backup policy.
    D. protected under the business impact analysis (BIA).

40. Which of the following MOST effectively ensures that service provider controls are within the guidelines set forth in the organization's information security policy?

    A. Service level monitoring
    B. Penetration testing
    C. Security awareness training
    D. Periodic auditing

41. Which of the following metrics is the MOST useful in measuring the monitoring of violation logs?

    A. Penetration attempts investigated
    B. Violation log reports produced
    C. Violation log entries
    D. Frequency of corrective actions taken

42. The person responsible for ensuring that information is classified is the:

    A. security manager.
    B. technology group.
    C. data owner.
    D. senior management.

SAMPLE EXAM

43. Management wants to ensure that IT is successful in delivering against business requirements. Which of the following **BEST** supports that effort?

   A. An internal control system or framework
   B. A cost-benefit analysis
   C. A return on investment (ROI) analysis
   D. A benchmark process

44. Which of the following is the **MOST** desirable strategy when developing risk mitigation options associated with the unavailability of IT services due to a natural disaster?

   A. Assume the worst-case incident scenarios.
   B. Target low-cost locations for alternate sites.
   C. Develop awareness focused on natural disasters.
   D. Enact multiple tiers of authority delegation.

45. Which of the following should be in place before a black box penetration test begins?

   A. A clearly stated definition of scope
   B. Previous test results
   C. Proper communication and awareness training
   D. An incident response plan

46. Which of the following is the **BEST** reason to perform a risk assessment?

   A. To satisfy regulatory requirements
   B. To budget appropriately for needed controls
   C. To analyze the effect on the business
   D. To help determine the current state of risk

47. Which of the following **MOST** enables risk-aware business decisions?

   A. Robust information security policies
   B. An exchange of accurate and timely information
   C. Skilled risk management personnel
   D. Effective process controls

48. Which of the following would data owners be **PRIMARILY** responsible for when establishing risk mitigation methods?

   A. Intrusion detection
   B. Antivirus controls
   C. User entitlement changes
   D. Platform security

49. Which of the following provides the **MOST** valuable input to incident response efforts?

   A. Qualitative analysis of threats
   B. The annual loss expectancy (ALE) total
   C. A vulnerability assessment
   D. Penetration testing

SAMPLE EXAM

50. Overall business risk for a particular threat can be expressed as the:

   A. magnitude of the impact should a threat source successfully exploit the vulnerability.
   B. likelihood of a given threat source exploiting a given vulnerability.
   C. product of the probability and magnitude of the impact if a threat exploits a vulnerability.
   D. collective judgment of the risk assessment team.

51. Previously accepted risk should be:

   A. reassessed periodically because the risk can be escalated to an unacceptable level due to revised conditions.
   B. removed from the risk log once it is accepted.
   C. accepted permanently because management has already spent resources (time and labor) to conclude that the risk level is acceptable.
   D. avoided next time because risk avoidance provides the best protection to the enterprise.

52. Which of the following is the MOST important information to include in a risk management strategic plan?

   A. Risk management staffing requirements
   B. The risk management mission statement
   C. Risk mitigation investment plans
   D. The current state and desired future state

53. An operations manager assigns monitoring responsibility of key risk indicators (KRIs) to line staff. Which of the following is MOST effective in validating the effort?

   A. Reported results should be independently reviewed.
   B. Line staff should complete risk management training.
   C. The threshold should be determined by risk management.
   D. Indicators should have benefits that exceed their costs.

54. Which of the following is the BEST indicator of high maturity of an enterprise's IT risk management process?

   A. People have appropriate awareness of risk and are comfortable talking about it.
   B. Top management is prepared to invest more money in IT security.
   C. Risk assessment is encouraged in all areas of IT and business management.
   D. Business and IT are aligned in risk assessment and risk ranking.

55. When using a formal approach to respond to a security-related incident, which of the following provides the GREATEST benefit from a legal perspective?

   A. Proving adherence to statutory audit requirements
   B. Proving adherence to corporate data protection requirements
   C. Demonstrating due care
   D. Working with law enforcement agencies

56. The GREATEST advantage of performing a business impact analysis (BIA) is that it:

   A. does not have to be updated because the impact will not change.
   B. promotes continuity awareness in the enterprise.
   C. can be performed using only qualitative estimates.
   D. eliminates the need to perform a risk analysis.

SAMPLE EXAM

57. Obtaining senior management commitment and support for information security investments can **BEST** be accomplished by a business case that:

   A. explains the technical risk to the enterprise.
   B. includes industry best practices as they relate to information security.
   C. details successful attacks against a competitor.
   D. ties security risk to organizational business objectives.

58. Which of the following is the **MOST** important factor when designing IS controls in a complex environment?

   A. Development methodologies
   B. Scalability of the solution
   C. Technical platform interfaces
   D. Stakeholder requirements

59. Which of the following uses risk scenarios when estimating the likelihood and impact of significant risk to the organization?

   A. An IT audit
   B. A security gap analysis
   C. A threat and vulnerability assessment
   D. An IT security assessment

60. Strong authentication is:

   A. an authentication technique formally approved by a standardization organization.
   B. the simultaneous use of several authentication techniques, e.g., password and badge.
   C. an authentication system that makes use of cryptography.
   D. an authentication system that uses biometric data to identify a person, i.e., a fingerprint.

61. An objective of a risk management program is to:

   A. maintain residual risk at an acceptable level.
   B. implement preventive controls for every threat.
   C. remove all inherent risk.
   D. reduce inherent risk to zero.

62. Which of the following **BEST** helps identify information systems control deficiencies?

   A. Gap analysis
   B. The current IT risk profile
   C. The IT controls framework
   D. Countermeasure analysis

63. An enterprise has learned of a security breach at another entity that utilizes similar technology. The **MOST** important action a risk practitioner should take is to:

   A. assess the likelihood of the incident occurring at the risk practitioner's enterprise.
   B. discontinue the use of the vulnerable technology.
   C. report to senior management that the enterprise is not affected.
   D. remind staff that no similar security breaches have taken place.

SAMPLE EXAM

64. The **PRIMARY** reason to report significant changes in IT risk to management is to:

    A. update the information asset inventory on a periodic basis.
    B. update the values of probability and impact for the related risk.
    C. reconsider the degree of importance of existing information assets.
    D. initiate an appropriate risk response for impacted information assets.

65. Which of the following is a behavior of risk avoidance?

    A. Take no action against the risk.
    B. Outsource the related process.
    C. Insure against a specific event.
    D. Exit the process that gives rise to risk.

66. Shortly after performing the annual review and revision of corporate policies, a risk practitioner becomes aware that a new law may affect security requirements for the human resources system. The risk practitioner should:

    A. analyze in detail how the law may affect the enterprise.
    B. ensure necessary adjustments are implemented during the next review cycle.
    C. initiate an *ad hoc* revision of the corporate policy.
    D. notify the system custodian to implement changes.

67. Which of the following is the **BEST** method to ensure the overall effectiveness of a risk management program?

    A. Assignment of risk within the enterprise
    B. Comparison of the program results with industry standards
    C. Participation by applicable members of the enterprise
    D. User assessment of changes in risk

68. The **PRIMARY** goal of a postincident review is to:

    A. gather evidence for subsequent legal action.
    B. identify ways to improve the response process.
    C. identify individuals who failed to take appropriate action.
    D. make a determination as to the identity of the attacker.

69. Who **MUST** give the final sign-off on the IT risk management plan?

    A. IT auditors performing the risk assessment
    B. Business process owners
    C. Senior management
    D. IT security administrators

70. The **PRIMARY** concern of a risk practitioner documenting a formal data retention policy is:

    A. storage availability.
    B. applicable organizational standards.
    C. generally accepted industry best practices.
    D. business requirements.

SAMPLE EXAM

71. A third party is engaged to develop a business application. Which of the following **BEST** measures for the existence of back doors?

    A. Security code reviews for the entire application
    B. System monitoring for traffic on network ports
    C. Reverse engineering the application binaries
    D. Running the application from a high-privileged account on a test system

72. A global enterprise that is subject to regulation by multiple governmental jurisdictions with differing requirements should:

    A. bring all locations into conformity with the aggregate requirements of all governmental jurisdictions.
    B. bring all locations into conformity with a generally accepted set of industry best practices.
    C. establish a baseline standard incorporating those requirements that all jurisdictions have in common.
    D. establish baseline standards for all locations and add supplemental standards as required.

73. Which of the following is the **BEST** indicator that incident response training is effective?

    A. Decreased reporting of security incidents to the incident response team
    B. Increased reporting of security incidents to the incident response team
    C. Decreased number of password resets
    D. Increased number of identified system vulnerabilities

74. Which of the following is **MOST** relevant to include in a cost-benefit analysis of a two-factor authentication system?

    A. The approved budget of the project
    B. The frequency of incidents
    C. The annual loss expectancy (ALE) of incidents
    D. The total cost of ownership (TCO)

75. During a risk management exercise, an analysis was conducted on the identified risk and mitigations were identified. Which choice **BEST** reflects residual risk?

    A. Risk left after the implementation of new or enhanced controls
    B. Risk mitigated as a result of the implementation of new or enhanced controls
    C. Risk identified prior to implementation of new or enhanced controls
    D. Risk classified as high after the implementation of new or enhanced controls

76. As part of fire drill testing, designated doors swing open, as planned, to allow employees to leave the building faster. An observer notices that this practice allows unauthorized personnel to enter the premises unnoticed. The **BEST** way to alter the process is to:

    A. stop the designated doors from opening automatically in case of a fire.
    B. include the local police force to guard the doors in case of fire.
    C. instruct the facilities department to guard the doors and have staff show their badge when exiting the building.
    D. assign designated personnel to guard the doors once the alarm sounds.

77. Which of the following is the **GREATEST** challenge of performing a quantitative risk analysis?

    A. Obtaining accurate figures on the impact of a realized threat
    B. Obtaining accurate figures on the value of assets
    C. Calculating the annualized loss expectancy (ALE) of a specific threat
    D. Obtaining accurate figures on the frequency of specific threats

SAMPLE EXAM

78. Which of the following approaches to corporate policy **BEST** supports an enterprise's expansion to other regions, where different local laws apply?

    A. A global policy that does not contain content that might be disputed at a local level
    B. A global policy that is locally amended to comply with local laws
    C. A global policy that complies with law at corporate headquarters and that all employees must follow
    D. Local policies to accommodate laws within each region

79. Which of the following factors will have the **GREATEST** impact on the type of information security governance model that an enterprise adopts?

    A. The number of employees
    B. The enterprise's budget
    C. The organizational structure
    D. The type of technology that the enterprise uses

80. Which of the following would **BEST** help an enterprise select an appropriate risk response?

    A. The degree of change in the risk environment
    B. An analysis of risk that can be transferred were it not eliminated
    C. The likelihood and impact of various risk scenarios
    D. An analysis of control costs and benefits

81. Which of the following situations is **BEST** addressed by transferring risk?

    A. An antiquated fire suppression system in the computer room
    B. The threat of disgruntled employee sabotage
    C. The possibility of the loss of a universal serial bus (USB) removable media drive
    D. A building located in a 100-year flood plain

82. A risk response report includes recommendations for:

    A. acceptance.
    B. assessment.
    C. evaluation.
    D. quantification.

83. Which of the following **BEST** provides message integrity, sender identity authentication and nonrepudiation?

    A. Symmetric cryptography
    B. Message hashing
    C. Message authentication code
    D. Public key infrastructure (PKI)

84. Which of the following is the **MOST** effective measure to protect data held on mobile computing devices?

    A. Protection of data being transmitted
    B. Encryption of stored data
    C. Power-on passwords
    D. Biometric access control

85. Which of the following will have the **MOST** significant impact on standard information security governance models?

    A. Number of employees
    B. Cultural differences between physical locations
    C. Complexity of the organizational structure
    D. Evolving legislative requirements

86. A new regulation for safeguarding information processed by a specific type of transaction has come to the attention of an IT manager. The manager should **FIRST**:

    A. meet with stakeholders to decide how to comply.
    B. analyze the key risk in the compliance process.
    C. update the existing security/privacy policy.
    D. assess whether existing controls meet the regulation.

87. One way to determine control effectiveness is by determining:

    A. the test results of intended objectives.
    B. whether it is preventive, detective or compensatory.
    C. the capability of providing notification of failure.
    D. the evaluation and analysis of reliability.

88. Which of the following environments typically represents the **GREATEST** risk to organizational security?

    A. An enterprise data warehouse
    B. A load-balanced, web server cluster
    C. A centrally managed data switch
    D. A locally managed file server

89. Which of the following reviews will provide the **MOST** insight into an enterprise's risk management capabilities?

    A. A capability maturity model (CMM) review
    B. A capability comparison with industry standards or regulations
    C. A self-assessment of capabilities
    D. An internal audit review of capabilities

90. Which of the following assessments of an enterprise's risk monitoring process will provide the **BEST** information about its alignment with industry-leading practices?

    A. A capability assessment by an outside firm
    B. A self-assessment of capabilities
    C. An independent benchmark of capabilities
    D. An internal audit review of capabilities

91. Which of the following is **MOST** important for measuring the effectiveness of a security awareness program?

    A. Increased interest in focus groups on security issues
    B. A reduced number of security violation reports
    C. A quantitative evaluation to ensure user comprehension
    D. An increased number of security violation reports

SAMPLE EXAM

92. When a significant vulnerability is discovered in the security of a critical web server, immediate notification should be made to the:

    A. development team to remediate.
    B. data owners to mitigate damage.
    C. system owner to take corrective action.
    D. incident response team to investigate.

93. Which of the following is MOST important when evaluating and assessing risk to an enterprise or business process?

    A. Identification of controls that are currently in place to mitigate identified risk
    B. Threat intelligence, including likelihood of identified threats
    C. Historical risk assessment data
    D. Control testing results

94. Which of the following is the MOST important requirement for setting up an information security infrastructure for a new system?

    A. Performing a business impact analysis (BIA)
    B. Considering personal devices as part of the security policy
    C. Basing the information security infrastructure on a risk assessment
    D. Initiating IT security training and familiarization

95. Which of the following will produce comprehensive results when performing a qualitative risk analysis?

    A. A vulnerability assessment
    B. Scenarios with threats and impacts
    C. The value of information assets
    D. Estimated productivity losses

96. Which of the following is the BEST way to verify that critical production servers are utilizing up-to-date antivirus signature files?

    A. Check a sample of servers.
    B. Verify the date that signature files were last pushed out.
    C. Use a recently identified benign virus to test whether it is quarantined.
    D. Research the most recent signature file, and compare it to the console.

97. Which of the following is the GREATEST benefit of a risk-aware culture?

    A. Issues are escalated when suspicious activity is noticed.
    B. Controls are double-checked to anticipate any issues.
    C. Individuals communicate with peers for knowledge sharing.
    D. Employees are self-motivated to learn about costs and benefits.

98. Risk assessment techniques should be used by a risk practitioner to:

    A. maximize the return on investment (ROI).
    B. provide documentation for auditors and regulators.
    C. justify the selection of risk mitigation strategies.
    D. quantify the risk that would otherwise be subjective.

SAMPLE EXAM

99. Which of the following choices will **BEST** protect the enterprise from financial risk?

    A. Insuring against the risk
    B. Updating the IT risk registry
    C. Improving staff training in the risk area
    D. Outsourcing the process to a third party

100. An enterprise has outsourced personnel data processing to a supplier, and a regulatory violation occurs during processing. Who will be held legally responsible?

    A. The supplier, because it has the operational responsibility
    B. The enterprise, because it owns the data
    C. The enterprise and the supplier
    D. The supplier, because it did not comply with the contract

101. Information security procedures should:

    A. be updated frequently as new software is released.
    B. underline the importance of security governance.
    C. define the allowable limits of behavior.
    D. describe security baselines for each platform.

102. Which of the following **BEST** addresses the risk of data leakage?

    A. Incident response procedures
    B. File backup procedures
    C. Acceptable use policies (AUPs)
    D. Database integrity checks

103. When performing a risk assessment on the impact of losing a server, calculating the monetary value of the server should be based on the:

    A. cost to obtain a replacement.
    B. annual loss expectancy (ALE).
    C. cost of the software stored.
    D. original cost to acquire.

104. A business impact analysis (BIA) is **PRIMARILY** used to:

    A. estimate the resources required to resume and return to normal operations after a disruption.
    B. evaluate the impact of a disruption to an enterprise's ability to operate over time.
    C. calculate the likelihood and impact of known threats on specific functions.
    D. evaluate high-level business requirements.

105. Which of the following is the **MOST** effective way to treat a risk such as a natural disaster that has a low probability and a high impact level?

    A. Eliminate the risk.
    B. Accept the risk.
    C. Transfer the risk.
    D. Implement countermeasures.

SAMPLE EXAM

106. A risk assessment process that uses likelihood and impact in calculating the level of risk is a:

    A. qualitative process.
    B. failure modes and effects analysis (FMEA).
    C. fault tree analysis.
    D. quantitative process.

107. Which of the following is the **MOST** important reason for conducting security awareness programs throughout an enterprise?

    A. Reducing the risk of a social engineering attack
    B. Training personnel in security incident response
    C. Informing business units about the security strategy
    D. Maintaining evidence of training records to ensure compliance

108. Which of the following **BEST** describes the risk-related roles and responsibilities of an organizational business unit (BU)? The BU management team:

    A. owns the mitigation plan for the risk belonging to their BU, while board members are responsible for identifying and assessing risk as well as reporting on that risk to the appropriate support functions.
    B. owns the risk and is responsible for identifying, assessing and mitigating risk as well as reporting on that risk to the appropriate support functions and the board of directors.
    C. carries out the respective risk-related responsibilities, but ultimate accountability for the day-to-day work of risk management and goal achievement belongs to the board members.
    D. is ultimately accountable for the day-to-day work of risk management and goal achievement, and board members own the risk.

109. The board of directors of a one-year-old start-up company has asked their chief information officer (CIO) to create all of the enterprise's IT policies and procedures. Which of the following should the CIO create **FIRST**?

    A. The strategic IT plan
    B. The data classification scheme
    C. The information architecture document
    D. The technology infrastructure plan

110. Which of the following is **MOST** critical when system configuration files for a critical enterprise application system are being reviewed?

    A. Configuration files are frequently changed.
    B. Changes to configuration files are recorded.
    C. Access to configuration files is not restricted.
    D. Configuration values do not impact system efficiency.

111. Where are key risk indicators (KRIs) **MOST** likely identified when initiating risk management across a range of projects?

    A. Risk governance
    B. Risk response
    C. Risk analysis
    D. Risk monitoring

SAMPLE EXAM

112. When proposing the implementation of a specific risk mitigation activity, a risk practitioner **PRIMARILY** utilizes a:

    A. technical evaluation report.
    B. business case.
    C. vulnerability assessment report.
    D. budgetary requirements.

113. Which of the following is the **MOST** appropriate metric to measure how well the information security function is managing the administration of user access?

    A. Elapsed time to suspend accounts of terminated users
    B. Elapsed time to suspend accounts of users transferring
    C. Ratio of actual accounts to actual end users
    D. Percent of accounts with configurations in compliance

114. Which of the following is the **BEST** option to ensure that corrective actions are taken after a risk assessment is performed?

    A. Conduct a follow-up review.
    B. Interview staff member(s) responsible for implementing the corrective action.
    C. Ensure that an organizational executive documents that the corrective action was taken.
    D. Run a monthly report and verify that the corrective action was taken.

115. Which of the following devices should be placed within a demilitarized zone (DMZ)?

    A. An authentication server
    B. A mail relay
    C. A firewall
    D. A router

116. Which of the following is the **BEST** metric to manage the information security program?

    A. The number of systems that are subject to intrusion detection
    B. The amount of downtime caused by security incidents
    C. The time lag between detection, reporting and acting on security incidents
    D. The number of recorded exceptions from the minimum information security requirements

117. Which of the following is the **BEST** way to ensure that a corporate network is adequately secured against external attack?

    A. Utilize an intrusion detection system (IDS).
    B. Establish minimum security baselines.
    C. Implement vendor recommended settings.
    D. Perform periodic penetration testing.

118. Which of the following **BEST** assists in the proper design of an effective key risk indicator (KRI)?

    A. Generating the frequency of reporting cycles to report on the risk
    B. Preparing a business case that includes the measurement criteria for the risk
    C. Conducting a risk assessment to provide an overview of the key risk
    D. Documenting the operational flow of the business from beginning to end

SAMPLE EXAM

119. Which of the following BEST describes the information needed for each risk on a risk register?

    A. Various risk scenarios with their date, description, impact, probability, risk score, mitigation action and owner
    B. Various risk scenarios with their date, description, risk score, cost to remediate, communication plan, and owner
    C. Various risk scenarios with their date, description, impact, cost to remediate, and owner
    D. Various activities leading to risk management planning

120. The MOST important objective of regularly testing information system controls is to:

    A. identify design flaws, failures and redundancies.
    B. provide the necessary evidence to support management assertions.
    C. assess the control risk and formulate an opinion on the level of reliability.
    D. evaluate the need for a risk assessment and indicate the corrective action(s) to be taken, where applicable.

121. Which of the following is MOST important for determining what security measures to put in place for a critical information system?

    A. The number of threats to the system
    B. The level of acceptable risk to the enterprise
    C. The number of vulnerabilities in the system
    D. The existing security budget

122. Business continuity plans (BCPs) should be written and maintained by:

    A. the information security and information technology functions.
    B. representatives from all functional units.
    C. the risk management function.
    D. executive management.

123. Which of the following will BEST prevent external security attacks?

    A. Securing and analyzing system access logs
    B. Network address translation
    C. Background checks for temporary employees
    D. Static Internet protocol (IP) addressing

124. Investments in risk management technologies should be based on:

    A. audit recommendations.
    B. vulnerability assessments.
    C. business climate.
    D. value analysis.

125. The MAIN objective of IT risk management is to:

    A. prevent loss of IT assets.
    B. provide timely management reports.
    C. ensure regulatory compliance.
    D. enable risk-aware business decisions.

SAMPLE EXAM

126. Implementing continuous monitoring controls is the **BEST** option when:

    A. legislation requires strong information security controls.
    B. incidents may have a high impact and frequency.
    C. incidents may have a high impact, but low frequency.
    D. electronic commerce (e-commerce) is a primary business driver.

127. Which of the following is the **FIRST** step when developing a risk monitoring program?

    A. Developing key indicators to monitor outcomes
    B. Gathering baseline data on indicators
    C. Analyzing and reporting findings
    D. Conducting a capability assessment

128. During a quarterly interdepartmental risk assessment, the IT operations center indicates a heavy increase of malware attacks. Which of the following recommendations to the business is **MOST** appropriate?

    A. Contract with a new anti-malware software vendor because the current solution seems ineffective.
    B. Close down the Internet connection to prevent employees from visiting infected web sites.
    C. Make the number of malware attacks part of each employee's performance metrics.
    D. Increase employee awareness training, including end-user roles and responsibilities.

129. An enterprise recently developed a breakthrough technology that could provide a significant competitive edge. Which of the following **FIRST** governs how this information is to be protected from within the enterprise?

    A. The data classification policy
    B. The acceptable use policy
    C. Encryption standards
    D. The access control policy

130. What is the **MOST** effective method to evaluate the potential impact of legal, regulatory and contractual requirements on business objectives?

    A. A compliance-oriented gap analysis
    B. Interviews with business process stakeholders
    C. A mapping of compliance requirements to policies and procedures
    D. A compliance-oriented business impact analysis (BIA)

131. Which of the following areas is **MOST** susceptible to the introduction of an information-security-related vulnerability?

    A. Tape backup management
    B. Database management
    C. Configuration management
    D. Incident response management

132. Which of the following is a control designed to prevent segregation of duties (SoD) violations?

    A. Enabling IT audit trails
    B. Implementing two-way authentication
    C. Reporting access log violations
    D. Implementing role-based access

133. Which of the following can be expected when a key control is being maintained at an optimal level?

    A. The shortest lead time until the control breach comes to the surface
    B. Balance between control effectiveness and cost
    C. An adequate maturity level of the risk management process
    D. An accurate estimation of operational risk amounts

134. Which of the following MOST likely indicates that a customer data warehouse should remain in-house rather than be outsourced to an offshore operation?

    A. The telecommunications costs may be much higher in the first year.
    B. Privacy laws may prevent a cross-border flow of information.
    C. Time zone differences may impede communications between IT teams.
    D. Software development may require more detailed specifications.

135. When configuring a biometric access control system that protects a high-security data center, the system's sensitivity level should be set to:

    A. a lower equal error rate (EER).
    B. a higher false acceptance rate (FAR).
    C. a higher false reject rate (FRR).
    D. the crossover error rate exactly.

136. Which of the following BEST indicates a successful risk management practice?

    A. Control risk is tied to business units.
    B. Overall risk is quantified.
    C. Residual risk is minimized.
    D. Inherent risk is eliminated.

137. The PRIMARY result of a risk management process is:

    A. a defined business plan.
    B. input for risk-aware decisions.
    C. data classification.
    D. minimized residual risk.

138. Which type of risk assessment methods involves conducting interviews and using anonymous questionnaires by subject matter experts?

    A. Quantitative
    B. Probabilistic
    C. Monte Carlo
    D. Qualitative

139. Acceptable risk for an enterprise is achieved when:

    A. transferred risk is minimized.
    B. control risk is minimized.
    C. inherent risk is minimized.
    D. residual risk is within tolerance levels.

SAMPLE EXAM

140. Malware has been detected that redirects users' computers to web sites crafted specifically for the purpose of fraud. The malware changes domain name system (DNS) server settings, redirecting users to sites under the hackers' control. This scenario BEST describes a:

    A. man-in-the-middle (MITM) attack.
    B. phishing attack.
    C. pharming attack.
    D. social engineering attack.

141. A global financial institution has decided not to take any further action on a denial of service (DoS) risk found by the risk assessment team. The MOST likely reason for making this decision is that:

    A. the needed countermeasure is too complicated to deploy.
    B. there are sufficient safeguards in place to prevent this risk from happening.
    C. the likelihood of the risk occurring is unknown.
    D. the cost of countermeasure outweighs the value of the asset and potential loss.

142. When developing risk scenarios for an enterprise, which of the following is the BEST approach?

    A. The top-down approach for capital-intensive enterprises
    B. The top-down approach because it achieves automatic buy-in
    C. The bottom-up approach for unionized enterprises
    D. The top-down and the bottom-up approach because they are complementary

143. Which of the following is the MOST prevalent risk in the development of end-user computing (EUC) applications?

    A. Increased development and maintenance costs
    B. Increased application development time
    C. Impaired decision making due to diminished responsiveness to requests for information
    D. Applications not subjected to testing and IT general controls

144. Which of the following practices BEST mitigates the risk associated with outsourcing a business function?

    A. Performing audits to verify compliance with contract requirements
    B. Requiring all vendor staff to attend annual awareness training sessions
    C. Retaining copies of all sensitive data on internal systems
    D. Reviewing the financial records of the vendor to verify financial soundness

145. Which of the following is the BEST approach when malicious code from a spear phishing attack resides on the network and the finance department is concerned that scanning the network will slow down work and delay quarter-end reporting?

    A. Instruct finance to finalize quarter-end reporting, then perform a scan of the entire network.
    B. Block all outgoing traffic to avoid outbound communication to the expecting command host.
    C. Scan network devices that are not supporting financial reporting, and then scan the critical finance drives at night.
    D. Perform a staff survey and ask staff to report if they are aware of the enterprise being a target of a spear phishing attack.

146. Which of the following should be of MOST concern to a risk practitioner?

    A. Failure to notify the public of an intrusion
    B. Failure to notify police of an attempted intrusion
    C. Failure to internally report a successful attack
    D. Failure to examine access rights periodically

SAMPLE EXAM

147. Assessing information systems risk is **BEST** achieved by:

    A. using the enterprise's past actual loss experience to determine current exposure.
    B. reviewing published loss statistics from comparable organizations.
    C. evaluating threats associated with existing information systems assets and information systems projects.
    D. reviewing information systems control weaknesses identified in audit reports.

148. A chief information security officer (CISO) has recommended several controls such as anti-malware to protect the enterprise's information systems. Which approach to handling risk is the CISO recommending?

    A. Risk transference
    B. Risk mitigation
    C. Risk acceptance
    D. Risk avoidance

149. As part of risk monitoring, the administrator of a two-factor authentication system identifies a trusted independent source indicating that the algorithm used for generating keys has been compromised. The vendor of the authentication system has not provided further information. Which of the following is the **BEST** initial course of action?

    A. Wait for the vendor to formally confirm the breach and provide a solution.
    B. Determine and implement suitable compensating controls.
    C. Identify all systems requiring two-factor authentication and notify their business owners.
    D. Disable the system and rely on the single-factor authentication until further information is received.

150. Which of the following is the **PRIMARY** reason that a risk practitioner determines the security boundary prior to conducting a risk assessment?

    A. To determine which laws and regulations apply
    B. To determine the scope of the risk assessment
    C. To determine the business owner(s) of the system
    D. To decide between conducting a quantitative or qualitative analysis

151. An enterprise has outsourced the majority of its IT department to a third party whose servers are in a foreign country. Which of the following is the **MOST** critical security consideration?

    A. A security breach notification may get delayed due to the time difference.
    B. Additional network intrusion detection sensors should be installed, resulting in additional cost.
    C. The enterprise could be unable to monitor compliance with its internal security and privacy guidelines.
    D. Laws and regulations of the country of origin may not be enforceable in the foreign country.

152. Which of the following is used to determine whether unauthorized modifications were made to production programs?

    A. An analytical review
    B. Compliance testing
    C. A system log analysis
    D. A forensic analysis

153. Who is **MOST** likely responsible for data classification?

    A. The data user
    B. The data owner
    C. The data custodian
    D. The system administrator

SAMPLE EXAM

154. Which of the following practices is MOST closely associated with risk monitoring?

   A. Assessment
   B. Mitigation
   C. Analysis
   D. Reporting

155. Which of the following is MOST essential for a risk management program to be effective?

   A. New risk detection
   B. A sound risk baseline
   C. Accurate risk reporting
   D. A flexible security budget

156. The MOST effective starting point to determine whether an IT system continues to meet the enterprise's business objectives is to conduct interviews with:

   A. executive management.
   B. IT management.
   C. business process owners.
   D. external auditors.

157. To determine the level of protection required for securing personally identifiable information, a risk practitioner should PRIMARILY consider the information:

   A. source.
   B. cost.
   C. sensitivity.
   D. validity.

158. Because of its importance to the business, an enterprise wants to quickly implement a technical solution that deviates from the company's policies. The risk practitioner should:

   A. recommend against implementation because it violates the company's policies.
   B. recommend revision of the current policy.
   C. conduct a risk assessment and allow or disallow based on the outcome.
   D. recommend a risk assessment and subsequent implementation only if residual risk is accepted.

159. Which of the following is the MAIN outcome of a business impact analysis (BIA)?

   A. Project prioritization
   B. Criticality of business processes
   C. The root cause of IT risk
   D. Third-party vendor risk

160. Which of the following is the PRIMARY reason for having the risk management process reviewed by independent risk auditors/assessors?

   A. To ensure that the risk results are consistent
   B. To ensure that the risk factors and risk profile are well defined
   C. To correct any mistakes in risk assessment
   D. To validate the control weaknesses for management reporting

SAMPLE EXAM

161. When transmitting personal information across networks, there MUST be adequate controls over:

   A. encrypting the personal information.
   B. obtaining consent to transfer personal information.
   C. ensuring the privacy of the personal information.
   D. change management.

162. Security technologies should be selected PRIMARILY on the basis of their:

   A. evaluation in security publications.
   B. compliance with industry standards.
   C. ability to mitigate risk to organizational objectives.
   D. cost compared to the enterprise's IT budget.

163. Which of the following is of MOST concern in a review of a virtual private network (VPN) implementation? Computers on the network are located:

   A. at the enterprise's remote offices.
   B. on the enterprise's internal network.
   C. at the backup site.
   D. in employees' homes.

164. A lack of adequate controls represents:

   A. a vulnerability.
   B. an impact.
   C. an asset.
   D. a threat.

165. Which of the following BEST ensures that information systems control deficiencies are appropriately remediated?

   A. A risk mitigation plan
   B. Risk reassessment
   C. Control risk reevaluation
   D. Countermeasure analysis

166. A network vulnerability assessment is intended to identify:

   A. security design flaws.
   B. zero-day vulnerabilities.
   C. misconfigurations and missing updates.
   D. malicious software and spyware.

167. Who should be accountable for the risk to an IT system that supports a critical business process?

   A. IT management
   B. Senior management
   C. The risk management department
   D. System users

SAMPLE EXAM

168. System backup and restore procedures can **BEST** be classified as:

   A. Technical controls
   B. Detective controls
   C. Corrective controls
   D. Deterrent controls

169. Which of the following is **MOST** important to determine when defining risk management strategies?

   A. Risk assessment criteria
   B. IT architecture complexity
   C. An enterprise disaster recovery plan (DRP)
   D. Organizational objectives and risk tolerance

170. The **MOST** effective method to conduct a risk assessment on an internal system in an organization is to start by understanding the:

   A. performance metrics and indicators.
   B. policies and standards.
   C. recent audit findings and recommendations.
   D. system and its subsystems.

171. As part of an enterprise risk management (ERM) program, a risk practitioner **BEST** leverages the work performed by an internal audit function by having it:

   A. design, implement and maintain the ERM process.
   B. manage and assess the overall risk awareness.
   C. evaluate ongoing changes to organizational risk factors.
   D. assist in monitoring, evaluating, examining and reporting on controls.

172. The **PRIMARY** reason for developing an enterprise security architecture is to:

   A. align security strategies between the functional areas of an enterprise and external entities.
   B. build a barrier between the IT systems of an enterprise and the outside world.
   C. help with understanding of the enterprise's technologies and the interactions between them.
   D. protect the enterprise from external threats and proactively monitor the corporate network.

173. Which of the following is the **BEST** method to analyze risk, incidents and related interdependencies to determine the impact on organizational goals?

   A. Security information and event management (SIEM) solutions
   B. A business impact analysis (BIA)
   C. Enterprise risk management (ERM) steering committee meetings
   D. Interviews with business leaders to develop a risk profile

174. An enterprise is hiring a consultant to help determine the maturity level of the risk management program. The **MOST** important element of the request for proposal (RFP) is the:

   A. sample deliverable.
   B. past experience of the engagement team.
   C. methodology used in the assessment.
   D. references from other organizations.

SAMPLE EXAM

175. A **PRIMARY** reason for initiating a policy exception process is when:

    A. the risk is justified by the benefit.
    B. policy compliance is difficult to enforce.
    C. operations are too busy to comply.
    D. users may initially be inconvenienced.

176. Security administration efforts are **BEST** reduced through the deployment of:

    A. access control lists (ACLs).
    B. discretionary access controls (DACs).
    C. mandatory access controls (MACs).
    D. role-based access controls (RBACs).

177. A substantive test to verify that tape library inventory records are accurate is:

    A. determining whether bar code readers are installed.
    B. conducting a physical count of the tape inventory.
    C. checking whether receipts and issues of tapes are accurately recorded.
    D. determining whether the movement of tapes is authorized.

178. It is **MOST** important for a risk evaluation to:

    A. take into account the potential size and likelihood of a loss.
    B. consider inherent and control risk.
    C. include a benchmark of similar companies in its scope.
    D. assume an equal degree of protection for all assets.

179. The annual expected loss of an asset—the annual loss expectancy (ALE)—is calculated as the:

    A. exposure factor (EF) multiplied by the annualized rate of occurrence (ARO).
    B. single loss expectancy (SLE) multiplied by the exposure factor (EF).
    C. single loss expectancy (SLE) multiplied by the annualized rate of occurrence (ARO).
    D. asset value (AV) multiplied by the single loss expectancy (SLE).

180. Which of the following techniques **BEST** helps determine whether there have been unauthorized program changes since the last authorized program update?

    A. A test data run
    B. An automated code comparison
    C. A code review
    D. A review of code migration procedures

181. In the risk management process, a cost-benefit analysis is **MAINLY** performed:

    A. as part of an initial risk assessment.
    B. as part of risk response planning.
    C. during an information asset valuation.
    D. when insurance is calculated for risk transfer.

SAMPLE EXAM

182. Despite a comprehensive security awareness program annually undertaken and assessed for all staff and contractors, an enterprise has experienced a breach through a spear phishing attack. What is the **MOST** effective way to improve security awareness?

   A. Review the security awareness program and improve coverage of social engineering threats.
   B. Launch a disciplinary process against the people who leaked the information.
   C. Perform a periodic social engineering test against all staff and communicate summary results to the staff.
   D. Implement a data loss prevention system that automatically points users to corporate policies.

183. Which of the following **BEST** describes the objective of a business impact analysis (BIA)?

   A. The identification of threats, risk and vulnerabilities that can adversely affect the enterprise
   B. The development of procedures for initial response and stabilization of situations during an emergency
   C. The identification of time-sensitive critical business functions and interdependencies
   D. The development of communication procedures in the case of a crisis impacting the business

184. Which of the following is the **BEST** way to ensure that an accurate risk register is maintained over time?

   A. Monitor key risk indicators (KRIs), and record the findings in the risk register.
   B. Publish the risk register centrally with workflow features that periodically poll risk assessors.
   C. Distribute the risk register to business process owners for review and updating.
   D. Utilize audit personnel to perform regular audits and to maintain the risk register.

185. Which of the following processes is **CRITICAL** for deciding prioritization of actions in a business continuity plan (BCP)?

   A. Risk assessment
   B. Vulnerability assessment
   C. A business impact analysis (BIA)
   D. Business process mapping

186. The **BEST** time to perform a penetration test is after:

   A. a high turnover in systems staff.
   B. an attempted penetration has occurred.
   C. various infrastructure changes are made.
   D. an audit has reported control weaknesses.

187. Risk management programs are designed to reduce risk to:

   A. the point at which the benefit exceeds the expense.
   B. a level that is too small to be measurable.
   C. a rate of return that equals the current cost of capital.
   D. a level that the enterprise is willing to accept.

188. Which of the following is the **PRIMARY** factor when deciding between conducting a quantitative or qualitative risk assessment?

   A. The corporate culture
   B. The amount of time available
   C. The availability of data
   D. The cost involved with risk assessment

SAMPLE EXAM

189. After the completion of a risk assessment, it is determined that the cost to mitigate the risk is much greater than the benefit to be derived. A risk practitioner should recommend to business management that the risk be:

   A. treated.
   B. terminated.
   C. accepted.
   D. transferred.

190. Which of the following is the **PRIMARY** reason for periodically monitoring key risk indicators (KRIs)?

   A. The cost of risk response needs to be minimized.
   B. Errors in results of KRIs need to be minimized.
   C. The risk profile may have changed.
   D. Risk assessment needs to be continually improved.

191. Which of the following is the **BEST** risk identification technique for an enterprise that allows employees to identify risk anonymously?

   A. The Delphi technique
   B. Isolated pilot groups
   C. A strengths, weaknesses, opportunities and threats (SWOT) analysis
   D. A root cause analysis

192. In a situation where the cost of anti-malware exceeds the loss expectancy of malware threats, what is the **MOST** viable risk response?

   A. Risk elimination
   B. Risk acceptance
   C. Risk transfer
   D. Risk mitigation

193. The **BEST** method for detecting and monitoring a hacker's activities without exposing information assets to unnecessary risk is to utilize:

   A. firewalls.
   B. bastion hosts.
   C. honeypots.
   D. screened subnets.

194. The preparation of a risk register begins in which risk management process?

   A. Risk response planning
   B. Risk monitoring and control
   C. Risk management planning
   D. Risk identification

195. Which of the following is the **MOST** effective way to ensure that third-party providers comply with the enterprise's information security policy?

   A. Security awareness training
   B. Penetration testing
   C. Service level monitoring
   D. Periodic auditing

SAMPLE EXAM

196. Which of the following is the **BEST** way to ensure that contract programmers comply with organizational security policies?

    A. Have the contractors acknowledge the security policies in writing.
    B. Perform periodic security reviews of the contractors.
    C. Explicitly refer to contractors in the security standards.
    D. Create penalties for noncompliance in the contracting agreement.

QUESTIONS 197 THROUGH 198 REFER TO THE FOLLOWING INFORMATION:

The chief information officer (CIO) of an enterprise has just received this year's IT security audit report. The report shows numerous open vulnerability findings on both business-critical and nonbusiness-critical information systems. The CIO briefed the chief executive officer (CEO) and board of directors on the findings and expressed his concern on the impact to the enterprise. He was informed that there are not enough funds to mitigate all of the findings from the report.

197. The CIO should respond to the findings identified in the IT security audit report by mitigating:

    A. the most critical findings on both the business-critical and nonbusiness-critical systems.
    B. all vulnerabilities on business-critical information systems first.
    C. the findings that are the least expensive to mitigate first to save funds.
    D. the findings that are the most expensive to mitigate first and leave all others until more funds become available.

198. Assuming that the CIO is unable to address all of the findings, how should the CIO deal with any findings that remain after available funds have been spent?

    A. Create a plan of actions and milestones for open vulnerabilities.
    B. Shut down the information systems with the open vulnerabilities.
    C. Reject the risk on the open vulnerabilities.
    D. Implement compensating controls on the systems with open vulnerabilities.

QUESTIONS 199 THROUGH 200 REFER TO THE FOLLOWING INFORMATION:

The board of directors of a one-year-old start-up company has asked their chief information officer (CIO) to create all of the enterprise's IT policies and procedures, which will be managed and approved by the IT steering committee. The IT steering committee will make all of the IT decisions for the enterprise, including those related to the technology budget.

199. The IT steering committee will be **BEST** represented by:

    A. members of the executive board.
    B. high-level members of the IT department.
    C. IT experts from outside of the enterprise.
    D. key members from each department.

200. Which type of IT organizational structure does the enterprise have?

    A. Project-based
    B. Centralized
    C. Decentralized
    D. Divisional

CRISC™ Review Questions, Answers & Explanations Manual 2013
SAMPLE EXAM ANSWER AND REFERENCE KEY

Exam Question #	Key	Ref. #	Exam Question #	Key	Ref. #	Exam Question #	Key	Ref. #	Exam Question #	Key	Ref. #
1	B	R4-11	51	A	R3-4	101	A	R5-3	151	D	R1-7
2	B	R1-19	52	D	R1-4	102	C	R4-7	152	B	R5-4
3	A	R1-12	53	A	R3-21	103	A	R1-18	153	B	R1-40
4	A	R5-36	54	A	R3-31	104	B	R4-21	154	D	R3-15
5	C	R5-32	55	C	R1-59	105	C	R2-8	155	A	R3-2
6	D	R1-36	56	B	R1-58	106	D	R1-32	156	C	R5-27
7	D	R3-25	57	D	R2-29	107	A	R1-27	157	C	R4-25
8	D	R4-4	58	D	R4-1	108	B	R1-56	158	D	R2-1
9	D	R3-1	59	C	R1-1	109	A	R1-62	159	B	R1-37
10	A	R5-11	60	B	R4-31	110	C	R5-31	160	B	R3-26
11	C	R4-8	61	A	R1-20	111	B	R3-19	161	C	R4-6
12	C	R4-27	62	A	R5-29	112	B	R2-2	162	C	R4-29
13	B	R1-35	63	A	R1-45	113	D	R3-29	163	D	R3-12
14	B	R4-24	64	D	R3-24	114	A	R5-24	164	A	R1-31
15	A	R1-57	65	D	R2-26	115	B	R4-10	165	A	R5-30
16	C	R3-34	66	A	R1-16	116	D	R5-20	166	C	R3-3
17	D	R4-22	67	C	R2-7	117	D	R5-8	167	B	R1-33
18	C	R2-10	68	B	R5-34	118	D	R3-33	168	C	R4-20
19	A	R3-5	69	C	R1-52	119	A	R1-55	169	D	R1-2
20	B	R4-17	70	D	R1-23	120	A	R5-26	170	D	R1-54
21	A	R4-30	71	A	R5-13	121	B	R2-28	171	D	R3-18
22	C	R4-14	72	D	R4-3	122	B	R4-18	172	A	R5-33
23	A	R4-32	73	B	R1-43	123	B	R4-13	173	B	R1-26
24	A	R1-38	74	D	R2-13	124	D	R4-2	174	C	R5-12
25	D	R2-18	75	A	R2-11	125	D	R1-47	175	A	R2-16
26	C	R1-61	76	D	R2-24	126	B	R5-5	176	D	R5-19
27	A	R1-39	77	D	R1-50	127	D	R3-10	177	B	R5-15
28	B	R2-31	78	B	R1-42	128	D	R2-25	178	A	R1-13
29	A	R3-28	79	C	R1-44	129	A	R1-10	179	C	R3-23
30	B	R2-21	80	D	R2-19	130	D	R1-14	180	B	R5-14
31	A	R5-35	81	D	R2-32	131	C	R1-25	181	B	R2-20
32	C	R3-27	82	A	R2-9	132	D	R4-19	182	C	R5-28
33	C	R3-20	83	D	R4-12	133	B	R3-22	183	C	R5-22
34	B	R1-30	84	B	R4-16	134	B	R3-9	184	B	R1-15
35	B	R3-14	85	C	R1-8	135	C	R4-15	185	C	R1-60
36	D	R5-25	86	D	R2-3	136	C	R3-6	186	C	R5-7
37	C	R4-26	87	A	R5-1	137	B	R2-23	187	D	R2-4
38	D	R2-5	88	D	R1-24	138	D	R3-13	188	C	R1-51
39	A	R1-6	89	A	R3-11	139	D	R2-30	189	C	R2-14
40	D	R5-23	90	C	R3-16	140	C	R1-11	190	C	R3-30
41	A	R5-6	91	D	R5-9	141	D	R2-12	191	A	R1-49
42	C	R4-5	92	C	R5-16	142	D	R1-29	192	B	R2-22
43	A	R4-28	93	B	R1-5	143	D	R3-17	193	C	R5-17
44	A	R1-48	94	C	R1-22	144	A	R1-34	194	D	R2-17
45	A	R5-10	95	B	R1-17	145	C	R5-21	195	D	R5-2
46	D	R1-3	96	A	R5-18	146	C	R3-8	196	B	R4-9
47	B	R3-7	97	A	R1-46	147	C	R1-21	197	B	R2-33
48	C	R1-9	98	C	R2-6	148	B	R2-27	198	A	R2-34
49	A	R1-41	99	A	R2-15	149	C	R3-32	199	D	R4-33
50	C	R1-28	100	B	R4-23	150	B	R1-53	200	B	R4-34

Reference example: RS5-9 = See domain 5, question 9 for explanation of the answer.

EVALUATION

ISACA continuously monitors the swift and profound professional, technological and environmental advances affecting risk practitioners. Recognizing these rapid advances, CRISC manuals are updated annually.

To assist ISACA in keeping abreast of these advances, please take a moment to evaluate the *CRISC™ Review Questions, Answers & Explanations Manual 2013*. Such feedback is valuable to fully serve the profession and future CRISC exam registrants.

To complete the evaluation on the web site, please go to *www.isaca.org/studyaidsevaluation*.

Thank you for your support and assistance.

Prepare for the 2013 CRISC Exams

2013 CRISC Review Resources for Exam Preparation and Professional Development

Successful Certified in Risk and Information Systems Control™ (CRISC™) exam candidates have an organized plan of study. To assist individuals with the development of a successful study plan, ISACA® offers several study aids and review courses to exam candidates. These include:

Study Aids

- *CRISC™ Review Manual 2013*

- *CRISC™ Review Questions, Answers & Explanations Manual 2013*

- *CRISC™ Review Questions, Answers & Explanations Manual 2013 Supplement*

To order, visit *www.isaca.org/criscbooks*.

Review Courses

- Chapter-sponsored review courses *(www.isaca.org/criscreview)*